BIOINFORMATICS

ALGORITHMS, CODING, DATA SCIENCE AND BIOSTATISTICS

4 BOOKS IN 1

BOOK 1
BIOINFORMATICS BASICS: AN INTRODUCTION TO ALGORITHMS AND CONCEPTS

BOOK 2
CODING IN BIOINFORMATICS: FROM SCRIPTING TO ADVANCED APPLICATIONS

BOOK 3
EXPLORING DATA SCIENCE IN BIOINFORMATICS: TECHNIQUES AND TOOLS FOR ANALYSIS

BOOK 4
MASTERING BIOSTATISTICS IN BIOINFORMATICS: ADVANCED METHODS AND APPLICATIONS

ROB BOTWRIGHT

Published by Rob Botwright
Library of Congress Cataloging-in-Publication Data
ISBN 978-1-83938-689-3
Cover design by Rizzo

Disclaimer

The contents of this book are based on extensive research and the best available historical sources. However, the author and publisher make no claims, promises, or guarantees about the accuracy, completeness, or adequacy of the information contained herein. The information in this book is provided on an "as is" basis, and the author and publisher disclaim any and all liability for any errors, omissions, or inaccuracies in the information or for any actions taken in reliance on such information. The opinions and views expressed in this book are those of the author and do not necessarily reflect the official policy or position of any organization or individual mentioned in this book. Any reference to specific people, places, or events is intended only to provide historical context and is not intended to defame or malign any group, individual, or entity. The information in this book is intended for educational and entertainment purposes only. It is not intended to be a substitute for professional advice or judgment. Readers are encouraged to conduct their own research and to seek professional advice where appropriate. Every effort has been made to obtain necessary permissions and acknowledgments for all images and other copyrighted material used in this book. Any errors or omissions in this regard are unintentional, and the author and publisher will correct them in future editions.

BOOK 1 - BIOINFORMATICS BASICS: AN INTRODUCTION TO ALGORITHMS AND CONCEPTS

BOOK 2 - CODING IN BIOINFORMATICS: FROM SCRIPTING TO ADVANCED APPLICATIONS

BOOK 3 - EXPLORING DATA SCIENCE IN BIOINFORMATICS: TECHNIQUES AND TOOLS FOR ANALYSIS

BOOK 4 - MASTERING BIOSTATISTICS IN BIOINFORMATICS: ADVANCED METHODS AND APPLICATIONS

Introduction

Welcome to the "Bioinformatics: Algorithms, Coding, Data Science, and Biostatistics" book bundle. This comprehensive collection of four books is designed to provide readers with a thorough understanding of bioinformatics – an interdisciplinary field that combines biology, computer science, statistics, and data science to analyze and interpret biological data.

Book 1, "Bioinformatics Basics: An Introduction to Algorithms and Concepts," serves as the cornerstone of this bundle. In this volume, readers will be introduced to fundamental concepts and algorithms in bioinformatics, including sequence analysis, sequence alignment, genetic variation, and the central dogma of molecular biology. By understanding these foundational principles, readers will gain insight into how computational methods are used to analyze biological data and solve real-world problems in the life sciences.

Moving on to Book 2, "Coding in Bioinformatics: From Scripting to Advanced Applications," readers will explore the practical implementation of bioinformatics algorithms and techniques. This volume covers scripting languages such as Python and R, as well as advanced applications in data manipulation, visualization, and machine learning. Through hands-on coding exercises and examples, readers will develop the skills necessary to write efficient and scalable code for bioinformatics analysis.

Book 3, "Exploring Data Science in Bioinformatics: Techniques and Tools for Analysis," shifts the focus to the burgeoning field of data science and its applications in bioinformatics. Here, readers will learn about exploratory data analysis, statistical inference, machine learning, and data visualization techniques specifically tailored for biological data. With a strong emphasis on practical applications, this book equips readers with the tools needed to extract meaningful insights from complex biological datasets.

Finally, Book 4, "Mastering Biostatistics in Bioinformatics: Advanced Methods and Applications," delves into the intricacies of biostatistics and its role in bioinformatics research. From advanced statistical methods to survival analysis and meta-analysis, readers will explore cutting-edge techniques for analyzing biological data and drawing meaningful conclusions from experimental studies.

Together, these four books provide a comprehensive overview of bioinformatics, from foundational concepts and coding skills to advanced data analysis and statistical methods. Whether you are a student, researcher, or practitioner in the life sciences, this book bundle offers valuable insights and practical knowledge to help you navigate the complexities of bioinformatics and advance your research in the field.

BOOK 1
BIOINFORMATICS BASICS
AN INTRODUCTION TO ALGORITHMS AND CONCEPTS

ROB BOTWRIGHT

Chapter 1: Introduction to Bioinformatics

Bioinformatics, at its core, represents the interdisciplinary field that merges biology, computer science, and information technology to analyze and interpret biological data. It serves as a pivotal bridge between biological sciences and computational methods, facilitating the understanding of complex biological systems through data-driven approaches. This chapter delves into the definition, scope, and significance of bioinformatics in modern scientific research.

Understanding Bioinformatics

In essence, bioinformatics encompasses a wide array of techniques and methodologies aimed at acquiring, processing, storing, analyzing, and interpreting biological data. This includes genetic sequences, protein structures, gene expression profiles, and various other molecular data types. By leveraging computational tools and algorithms, researchers in bioinformatics strive to extract meaningful insights from these vast datasets, ultimately advancing our understanding of biological phenomena.

Scope of Bioinformatics

The scope of bioinformatics is vast and continually expanding, driven by advancements in technology and the ever-increasing volume of biological data generated by various high-throughput experimental techniques. Key areas within the scope of bioinformatics include:

1. Genomics: Genomics focuses on the study of an organism's entire genome, including its structure, function, and evolution. Bioinformatics plays a crucial role in genome sequencing, assembly, annotation, and comparative genomics analysis. Tools such as BLAST (Basic Local Alignment Search Tool) are commonly used for sequence similarity searches, aiding in the identification of homologous genes across different species.

2. Proteomics: Proteomics involves the study of an organism's proteome, encompassing all of its proteins and their functions. Bioinformatics techniques are employed for protein sequence analysis, structure prediction, and functional annotation. Software packages like SWISS-MODEL and Phyre2 facilitate protein structure prediction based on homology modeling and threading approaches.

3. Transcriptomics: Transcriptomics focuses on the study of an organism's transcriptome, comprising all of its RNA transcripts, including mRNA, non-coding RNA, and splice variants. Bioinformatics tools enable the analysis of gene expression patterns, alternative splicing events, and regulatory networks. Popular tools such as DESeq2 and edgeR are utilized for differential gene expression analysis from RNA-seq data.

4. Metabolomics: Metabolomics involves the comprehensive analysis of small-molecule metabolites present within biological systems. Bioinformatics methods are employed for metabolite identification, quantification, and metabolic pathway analysis. Software tools like MetaboAnalyst and MZmine aid in the processing and interpretation of mass spectrometry data for metabolomics studies.

5. Systems Biology: Systems biology aims to understand biological systems as integrated networks of genes, proteins, and metabolites, rather than isolated components. Bioinformatics techniques are used to model and simulate complex biological systems, enabling the prediction of system behavior under different conditions. Platforms such as COPASI and CellDesigner facilitate the construction and analysis of biochemical network models.

Significance of Bioinformatics

The significance of bioinformatics in modern scientific research cannot be overstated. It has revolutionized various fields within biology and medicine, facilitating breakthroughs in drug discovery, disease diagnosis, personalized medicine, and agricultural biotechnology. By harnessing the power of computational analysis, bioinformatics accelerates the pace of biological discovery and innovation.

Deployment of Bioinformatics Techniques

To deploy bioinformatics techniques effectively, researchers typically rely on a combination of command-line tools, scripting languages, and specialized software packages. For instance, when analyzing sequencing data, researchers may use the following CLI command to align sequencing reads to a reference genome:

bashCopy code

```
bowtie2 -x <reference_index> -U <input_reads.fastq> -S <output_alignment.sam>
```

Similarly, for differential gene expression analysis from RNA-seq data, researchers may employ the following R script using the DESeq2 package:

RCopy code

```
library(DESeq2) countData <- read.csv("counts.csv", header = TRUE) metadata <- read.csv("metadata.csv", header = TRUE) dds <- DESeqDataSetFromMatrix(countData = countData,
```

colData = metadata, design = ~condition) dds <-
DESeq(dds) results <- results(dds)

By mastering these tools and techniques, researchers can effectively navigate the complex landscape of biological data analysis and contribute to advancements in various domains of life sciences.

In summary, bioinformatics serves as an indispensable tool for deciphering the intricacies of biological systems, from the molecular level to entire ecosystems. Its interdisciplinary nature and wide-ranging applications make it a cornerstone of modern scientific research, with profound implications for fields such as medicine, agriculture, and environmental science.

Historical Development and Milestones

The historical development of bioinformatics is a fascinating journey that spans several decades, marked by key milestones and transformative breakthroughs. This chapter explores the evolution of bioinformatics from its nascent beginnings to its current status as a cornerstone of modern biological research.

Early Beginnings

The roots of bioinformatics can be traced back to the mid-20th century when scientists began to explore the computational analysis of biological data. One of the earliest milestones in this journey was the development of computational methods for sequence alignment. In 1965, Margaret Dayhoff pioneered the field of computational biology by creating the first comprehensive database of protein sequences, known as the "Atlas of Protein Sequence and Structure." This monumental effort laid the foundation for subsequent advancements in sequence analysis and homology modeling.

The Genomic Era

The advent of high-throughput DNA sequencing technologies in the 1970s and 1980s ushered in the genomic era, revolutionizing the field of molecular biology. Frederick Sanger's groundbreaking work on DNA sequencing techniques paved the way for the sequencing of the first complete genome of a bacteriophage in 1977. This achievement marked the beginning of a new era in biological research, as scientists gained unprecedented access to the genetic blueprint of organisms.

GenBank and the Human Genome Project

In 1982, the National Institutes of Health (NIH) launched GenBank, a publicly accessible database for storing DNA sequence data. GenBank played a pivotal role in facilitating data sharing and collaboration among researchers worldwide, laying the groundwork for large-scale genome sequencing projects. One of the most ambitious endeavors in this regard was the Human Genome Project, initiated in 1990 with the goal of sequencing the entire human genome. The completion of the Human Genome Project in 2003 marked a historic milestone in bioinformatics, providing invaluable insights into human genetics and paving the way for personalized medicine.

Bioinformatics Algorithms and Tools

Throughout the 1990s and early 2000s, bioinformatics witnessed a proliferation of algorithms and software tools designed to analyze and interpret biological data. One notable advancement was the development of the Basic Local Alignment Search Tool (BLAST) by Altschul et al. in

1990. BLAST revolutionized sequence similarity searching, enabling researchers to identify homologous sequences in large databases with remarkable speed and accuracy. Another significant development was the creation of the Ensembl genome browser in 1999, providing a comprehensive platform for visualizing and analyzing genomic data from diverse species.

Next-Generation Sequencing

The emergence of next-generation sequencing (NGS) technologies in the late 2000s marked a paradigm shift in genomic research, enabling rapid and cost-effective sequencing of entire genomes. Platforms such as Illumina, Ion Torrent, and Pacific Biosciences revolutionized the field of genomics, generating massive volumes of sequencing data at unprecedented scale. Bioinformatics played a crucial role in processing and analyzing NGS data, driving advancements in fields such as personalized medicine, evolutionary biology, and agricultural genomics.

Omics Revolution

The past decade has witnessed the rise of the "omics" revolution, characterized by the integration of genomics, transcriptomics, proteomics, metabolomics, and other high-throughput data types. This multi-omics approach has enabled researchers to gain comprehensive insights into the molecular mechanisms underlying biological processes and disease states. Bioinformatics tools and techniques have played a central role in integrating and analyzing multi-omics data, facilitating systems-level understanding of complex biological systems.

Future Perspectives

Looking ahead, the field of bioinformatics is poised for continued growth and innovation, driven by advancements in technology, data science, and computational methods. Emerging trends such as single-cell sequencing, spatial transcriptomics, and artificial intelligence promise to further expand the frontiers of biological research. As we embark on this journey into the future, it is essential to recognize and appreciate the rich history and legacy of bioinformatics, which continues to shape the course of scientific discovery and innovation.

In summary, the historical development of bioinformatics is a testament to the ingenuity and perseverance of scientists who have pushed the boundaries of knowledge and transformed our understanding of the natural world. From humble beginnings to cutting-edge technologies, bioinformatics has evolved into a powerful interdisciplinary field with profound implications for biology, medicine, and beyond.

Chapter 2: Fundamentals of Molecular Biology

The discovery of the structure of DNA stands as one of the most significant milestones in the history of science. This chapter delves into the intricacies of DNA structure and its fundamental role in encoding genetic information, providing a comprehensive understanding of this essential molecule.

Understanding DNA Structure

Deoxyribonucleic acid, or DNA, is a long, double-stranded molecule that carries the genetic instructions necessary for the development, functioning, growth, and reproduction of all known living organisms and many viruses. The structure of DNA was elucidated by James Watson and Francis Crick in 1953, based on X-ray diffraction data collected by Rosalind Franklin and Maurice Wilkins. The iconic double helix structure of DNA consists of two polynucleotide chains twisted around each other in a spiral configuration, forming a ladder-like structure with complementary base pairs.

The Double Helix Model

At the heart of the DNA double helix are nucleotides, the building blocks of DNA. Each nucleotide comprises three components: a phosphate group, a sugar molecule (deoxyribose), and one of four nitrogenous bases— adenine (A), cytosine (C), guanine (G), or thymine (T). The structure of DNA is stabilized by hydrogen bonds between complementary base pairs: adenine pairs with thymine (A-T), and cytosine pairs with guanine (C-G). This complementary base pairing ensures the faithful replication and transmission of genetic information during cell division and DNA synthesis.

Major and Minor Grooves

The double helix structure of DNA features two distinct grooves: the major groove and the minor groove. These grooves result from the helical twisting of the DNA strands and provide access points for DNA-binding proteins and other molecules involved in gene expression, DNA replication, and repair processes. The major groove, wider and more accessible than the minor groove, serves as a primary site for protein-DNA interactions and regulatory protein binding.

Functions of DNA

DNA serves as the blueprint for life, encoding the instructions required for the synthesis of proteins, the molecular machines that carry out most cellular functions. The genetic information encoded in DNA is transcribed into messenger RNA (mRNA) molecules through a process called transcription. These mRNA molecules are then translated into proteins by ribosomes, the cellular machinery responsible for protein synthesis. The sequence of nucleotides in DNA determines the sequence of amino acids in proteins, thereby dictating the structure and function of proteins and ultimately shaping the phenotype of an organism.

Deploying DNA Analysis Techniques

Various techniques are employed to analyze DNA structure and function, providing insights into genetic variation, gene expression patterns, and regulatory mechanisms. One such technique is polymerase chain reaction (PCR), a method used to amplify specific DNA sequences. The following CLI command illustrates the PCR process:

bashCopy code

```
PCR    -i    <input_DNA_sequence.fasta>    -p
<primer_sequences.fasta>    -o
<output_amplified_products.fasta>
```

Another widely used technique is DNA sequencing, which allows for the determination of the nucleotide sequence of DNA molecules. Next-generation sequencing (NGS) platforms, such as Illumina and Oxford Nanopore, have revolutionized DNA sequencing, enabling high-throughput and cost-effective analysis of entire genomes. The following CLI command demonstrates the use of the popular sequence alignment tool, BWA (Burrows-Wheeler Aligner), for aligning sequencing reads to a reference genome:

bashCopy code

```
bwa    mem    -t    <number_of_threads>
<reference_genome.fasta>    <input_reads.fastq>    >
<output_alignment.sam>
```

Additionally, bioinformatics tools and software packages are utilized to analyze and interpret DNA sequence data, facilitating genome annotation, variant calling, and comparative genomics analysis. For instance, the following Python script utilizes the Biopython library to parse and analyze DNA sequences:

pythonCopy code

```
from Bio import SeqIO # Read input DNA sequence file
sequences    =    SeqIO.parse("input_sequence.fasta",
"fasta") # Iterate over sequences and compute GC
content for sequence in sequences: gc_content =
```

(sequence.seq.count("G") + sequence.seq.count("C")) /
len(sequence.seq) print(f"GC content of {sequence.id}:
{gc_content:.2f}")

By deploying these techniques and tools, researchers can unravel the mysteries of DNA structure and function, unlocking insights into the molecular basis of life and disease.

Central Dogma of Molecular Biology

The Central Dogma of Molecular Biology represents a foundational principle that governs the flow of genetic information within living organisms. This chapter delves into the intricacies of the Central Dogma, elucidating its significance in understanding the molecular mechanisms underlying life processes.

Overview of the Central Dogma

Proposed by Francis Crick in 1958, the Central Dogma of Molecular Biology postulates the flow of genetic information within cells, outlining the sequential processes of replication, transcription, and translation. At its core, the Central Dogma describes how genetic information encoded in DNA is transcribed into RNA and subsequently translated into proteins, the functional molecules that drive cellular processes.

Replication: DNA to DNA

The first step in the Central Dogma is DNA replication, wherein the genetic information stored in DNA is faithfully copied to produce an identical DNA molecule. DNA replication occurs prior to cell division, ensuring that each daughter cell receives a complete set of genetic instructions. The replication process involves unwinding of the DNA double helix, synthesis of complementary DNA strands by DNA polymerase enzymes, and proofreading mechanisms to maintain genomic integrity.

Deploying DNA Replication Techniques

To study DNA replication, researchers often utilize techniques such as polymerase chain reaction (PCR) to amplify specific DNA sequences. The following CLI command demonstrates the PCR process:

bashCopy code

```
PCR     -i     <input_DNA_sequence.fasta>     -p
<primer_sequences.fasta>                       -o
<output_amplified_products.fasta>
```

Alternatively, techniques such as DNA labeling with radioactive isotopes or fluorescent probes can be employed to visualize newly synthesized DNA strands during replication.

Transcription: DNA to RNA

The next step in the Central Dogma is transcription, wherein the genetic information stored in DNA is transcribed into RNA molecules. Transcription is carried out by RNA polymerase enzymes, which catalyze the synthesis of RNA strands complementary to the DNA template strand. The resulting RNA molecules, known as messenger RNA (mRNA), serve as templates for protein synthesis during the process of translation.

Deploying Transcription Techniques

To analyze transcriptional activity, researchers often employ techniques such as reverse transcription polymerase chain reaction (RT-PCR) to quantify mRNA expression levels. The following CLI command illustrates the RT-PCR process:

bashCopy code

RT-PCR   -i   <input_RNA_sample.fasta>   -o
<output_quantified_mRNA.fasta>
Alternatively, RNA sequencing (RNA-seq) can be used to comprehensively profile gene expression patterns across the entire transcriptome.

Translation: RNA to Protein
The final step in the Central Dogma is translation, wherein the genetic information encoded in mRNA is translated into protein sequences. Translation occurs on ribosomes, cellular organelles composed of ribosomal RNA (rRNA) and protein subunits. Transfer RNA (tRNA) molecules ferry amino acids to the ribosome, where they are assembled into polypeptide chains according to the sequence of codons on the mRNA template.

Deploying Translation Techniques
To study translation, researchers often use techniques such as ribosome profiling to monitor the translation efficiency of mRNA molecules. The following CLI command demonstrates the ribosome profiling process:
bashCopy code
ribosome_profiling   -i
<input_translated_mRNA_sequences.fasta>   -o
<output_ribosome_bound_mRNA.fasta>
Additionally, techniques such as mass spectrometry can be employed to identify and characterize proteins synthesized within cells.

Significance of the Central Dogma
The Central Dogma of Molecular Biology serves as a guiding principle for understanding the flow of genetic information and the molecular basis of life processes. By elucidating the sequential steps of replication,

transcription, and translation, the Central Dogma provides a framework for unraveling the complexities of cellular function and genetic regulation. Moreover, deviations from the Central Dogma, such as reverse transcription in retroviruses or RNA-based gene regulation mechanisms, highlight the dynamic nature of molecular processes within living organisms.

Future Perspectives

As advancements in technology and molecular biology continue to accelerate, the Central Dogma remains a cornerstone of scientific inquiry, guiding research efforts aimed at unraveling the mysteries of the genome, transcriptome, and proteome. By leveraging cutting-edge techniques and interdisciplinary approaches, researchers can further elucidate the intricacies of genetic information flow and its role in health, disease, and evolution. In summary, the Central Dogma of Molecular Biology represents a fundamental principle that underpins our understanding of genetic information transfer within cells. By elucidating the sequential processes of DNA replication, transcription, and translation, the Central Dogma provides a framework for exploring the molecular mechanisms underlying life processes and genetic regulation.

Chapter 3: Basics of Sequence Analysis

In the realm of bioinformatics, sequence data lies at the heart of genetic analysis, facilitating insights into the structure, function, and evolution of biological molecules. This chapter delves into the diverse types of sequence data and the formats commonly used to represent them, providing a comprehensive overview of this foundational aspect of bioinformatic analysis.

Types of Sequence Data

Sequence data encompasses a wide array of biological molecules, including DNA, RNA, and proteins, each with distinct roles and functions within living organisms. Understanding the different types of sequence data is essential for interpreting genetic information and unraveling the complexities of biological systems.

DNA Sequences

DNA sequences represent the genetic blueprint of an organism, encoding the instructions necessary for its development, functioning, and inheritance. DNA sequences are composed of four nucleotide bases: adenine (A), cytosine (C), guanine (G), and thymine (T). The linear arrangement of these nucleotides forms the unique genetic code of an organism, dictating its traits and characteristics.

RNA Sequences

RNA sequences play crucial roles in gene expression, serving as intermediaries between DNA and proteins. There are several types of RNA molecules, including messenger RNA (mRNA), transfer RNA (tRNA), and ribosomal RNA (rRNA), each with distinct functions in the cellular machinery. RNA sequences are also composed of nucleotide bases, but they contain uracil (U) instead of thymine (T) as found in DNA.

Protein Sequences

Protein sequences represent the amino acid chains that form the building blocks of proteins, the molecular machines that carry out most cellular functions. Proteins are composed of 20 different amino acids, each encoded by a specific triplet of nucleotides called a codon. The sequence of amino acids in a protein determines its structure, function, and interactions within the cell.

Common Sequence Data Formats

In bioinformatics, sequence data is typically represented and stored in various file formats, each designed to accommodate different types of sequence data and associated metadata. Understanding these formats is essential for processing, analyzing, and sharing sequence data effectively.

FASTA Format

The FASTA format is one of the most commonly used formats for representing nucleotide and protein sequences. In the FASTA format, each sequence is represented by a header line, preceded by a greater-than symbol (>), followed by one or more lines containing the sequence data. For example, a DNA sequence in FASTA format might look like this:

```
shellCopy code
>sequence_name
ATCGATCGATCGATCGATCGATCGATCGATCG
```

FASTQ Format

The FASTQ format is used to represent sequencing reads generated by high-throughput sequencing technologies, such as Illumina and Ion Torrent. In the FASTQ format, each sequence read is represented by four lines: a sequence identifier, the raw sequence data, a plus sign (+) separator, and quality scores corresponding to each base in the sequence. For example:

cssCopy code

```
@read1    ATCGATCGATCGATCGATCGATCGATCG    +
BBBBBBBBBBBBBBBBBBBBBBBBBBBBBBB
```

SAM/BAM Formats

The Sequence Alignment/Map (SAM) format and its binary counterpart, Binary Alignment/Map (BAM), are used to represent sequence alignment data, including mapped reads and their alignments to a reference genome. SAM files contain header lines followed by alignment records, each describing the alignment of a single sequence read to the reference genome. BAM files are binary versions of SAM files, optimized for efficient storage and retrieval.

Deploying Sequence Data Analysis Techniques

To analyze sequence data, researchers often use bioinformatics tools and software packages designed to process, align, and interpret sequence data. For instance, to align sequencing reads to a reference genome and generate a SAM file, researchers may use the Bowtie2 aligner with the following CLI command:

bashCopy code

```
bowtie2 -x <reference_index> -U <input_reads.fastq> -S <output_alignment.sam>
```

Subsequently, the SAM file can be converted to the BAM format and sorted using tools such as SAMtools:

bashCopy code

```
samtools    view    -b    <input_alignment.sam>    -o
<output_alignment.bam>    samtools    sort
<input_alignment.bam> -o <sorted_alignment.bam>
```
Additionally, bioinformatics software packages like Biopython and Bioconductor provide programming interfaces and libraries for working with sequence data in various formats, enabling custom analysis and visualization pipelines.

Significance of Sequence Data

Sequence data lies at the intersection of biology, computer science, and statistics, serving as the foundation for a wide range of bioinformatic analyses and applications. From genome assembly and annotation to comparative genomics and protein structure prediction, sequence data fuels discoveries in fields such as genetics, evolutionary biology, and personalized medicine. By leveraging advanced sequencing technologies and bioinformatics tools, researchers can unravel the complexities of the genetic code and gain insights into the mechanisms of life itself.

In summary, sequence data types and formats form the cornerstone of bioinformatics, enabling the analysis and interpretation of genetic information across diverse organisms and biological processes. By understanding the principles of sequence data representation and analysis, researchers can unlock the mysteries of the genome and harness the power of genetic information for scientific discovery and innovation.

Sequence Alignment: Concepts and Importance

Sequence alignment stands as a fundamental technique in bioinformatics, essential for comparing and analyzing biological sequences to unveil patterns, relationships, and evolutionary histories. This chapter delves into the concepts and significance of sequence alignment, elucidating its critical role in understanding genetic information and biological processes.

Understanding Sequence Alignment

At its core, sequence alignment involves the comparison of two or more sequences to identify regions of similarity, dissimilarity, and conserved motifs. The goal of sequence alignment is to infer evolutionary relationships, detect functional domains, and elucidate sequence variations across different organisms or within the same organism over time.

Pairwise Sequence Alignment

Pairwise sequence alignment is the simplest form of sequence alignment, involving the comparison of two sequences to identify regions of similarity or homology. One of the most widely used algorithms for pairwise sequence alignment is the Needleman-Wunsch algorithm, which employs dynamic programming to find the optimal alignment between two sequences based on a scoring matrix.

Deploying Pairwise Sequence Alignment Techniques

To perform pairwise sequence alignment using the Needleman-Wunsch algorithm, researchers can utilize bioinformatics tools such as EMBOSS or NCBI BLAST. The following CLI command demonstrates the use of NCBI BLAST for pairwise sequence alignment:

bashCopy code

```
blastn -query sequence1.fasta -subject sequence2.fasta -out alignment_results.txt
```

This command compares the sequences in the "sequence1.fasta" and "sequence2.fasta" files and outputs the alignment results to a text file named "alignment_results.txt".

Multiple Sequence Alignment

Multiple sequence alignment extends pairwise alignment to compare three or more sequences simultaneously, allowing for the identification of conserved regions and evolutionary patterns across a group of related sequences. Multiple sequence alignment is particularly useful for phylogenetic analysis, functional annotation, and structure prediction studies.

Deploying Multiple Sequence Alignment Techniques

To perform multiple sequence alignment, researchers often utilize software tools such as Clustal Omega or MUSCLE. The following CLI command demonstrates the use of Clustal Omega for multiple sequence alignment:

bashCopy code

```
clustalo          -i          input_sequences.fasta          -o
output_alignment.fasta
```

This command aligns the sequences in the "input_sequences.fasta" file and outputs the alignment results to a file named "output_alignment.fasta" using the Clustal Omega algorithm.

Importance of Sequence Alignment

Sequence alignment plays a pivotal role in various areas of biological research, contributing to our understanding of genome evolution, protein structure-function relationships, and disease mechanisms. Some key aspects of its importance include:

Evolutionary Analysis: Sequence alignment enables the reconstruction of evolutionary histories and the identification of conserved and variable regions within genomes and protein sequences. By aligning sequences from different species or homologous genes, researchers can infer evolutionary relationships and trace the origins of genetic diversity.

Functional Annotation: Sequence alignment aids in the annotation of functional domains, motifs, and regulatory elements within genomes and protein sequences. By aligning sequences with known functional annotations or structural domains, researchers can predict the functions of newly discovered genes and proteins, guiding experimental studies and functional characterization efforts.

Structure Prediction: Sequence alignment serves as a crucial step in protein structure prediction, facilitating the identification of conserved residues, secondary structure elements, and protein-protein interaction interfaces. By aligning protein sequences with experimentally determined structures or homologous models, researchers can generate hypotheses about protein structure and function, guiding drug discovery and protein engineering efforts.

Disease Genomics: Sequence alignment plays a vital role in disease genomics, enabling the identification of disease-causing mutations, genetic variations, and regulatory elements within genomes. By aligning sequences from healthy and diseased individuals, researchers can pinpoint genetic differences associated with disease susceptibility, progression, and response to treatment, paving the way for personalized medicine and precision therapeutics.

Future Perspectives

As sequencing technologies continue to advance and genomic datasets grow in size and complexity, sequence alignment remains a cornerstone of bioinformatics research, driving discoveries in fields such as comparative genomics, functional genomics, and systems biology. By integrating advanced algorithms, machine learning techniques, and high-performance computing resources, researchers can further enhance the accuracy, scalability, and efficiency of sequence alignment methods, unlocking new insights into the molecular basis of life and disease.

In summary, sequence alignment stands as a fundamental technique in bioinformatics, essential for comparing, analyzing, and interpreting biological sequences across diverse organisms and contexts. By understanding the concepts and importance of sequence alignment, researchers can unravel the mysteries of the genome, elucidate evolutionary relationships, and harness the power of genetic information for scientific discovery and innovation.

Chapter 4: Understanding Genetic Variation

Genetic variation serves as the foundation of biological diversity, driving evolution, adaptation, and inheritance across populations and species. This chapter explores the diverse types and sources of genetic variation, shedding light on the mechanisms that underlie genetic diversity within and between organisms.

Understanding Genetic Variation

Genetic variation refers to the differences in DNA sequences, gene expression patterns, and phenotypic traits observed among individuals within a population or species. These variations arise from mutations, genetic recombination, and other molecular mechanisms that introduce changes to the genetic material, shaping the diversity of life forms on Earth.

Types of Genetic Variation

Genetic variation manifests in various forms, each contributing to the complexity and adaptability of living organisms. Some key types of genetic variation include:

Single Nucleotide Polymorphisms (SNPs): SNPs are the most common type of genetic variation, involving single nucleotide substitutions at specific positions in the genome. SNPs can influence traits such as disease susceptibility, drug response, and physical characteristics, making them valuable markers for genetic studies and personalized medicine.

Insertions and Deletions (Indels): Indels refer to the insertion or deletion of nucleotides within the genome, leading to changes in gene structure and function.

Indels can cause frameshift mutations, altering the reading frame of genes and potentially disrupting protein synthesis.

Copy Number Variations (CNVs): CNVs involve changes in the number of copies of specific DNA segments within the genome, resulting from duplications, deletions, or rearrangements of genomic regions. CNVs can influence gene dosage, expression levels, and phenotypic traits, contributing to genetic diversity and disease susceptibility.

Structural Variations: Structural variations encompass large-scale alterations to the genomic architecture, including inversions, translocations, and chromosomal rearrangements. These structural variations can disrupt gene regulation, chromosome stability, and genomic integrity, leading to diseases such as cancer and developmental disorders.

Sources of Genetic Variation

Genetic variation arises from a multitude of sources, reflecting the interplay between evolutionary processes, environmental factors, and molecular mechanisms that shape the genetic landscape of populations. Some key sources of genetic variation include:

Mutation: Mutations represent the ultimate source of genetic variation, introducing changes to DNA sequences through errors in DNA replication, repair, or recombination. Mutations can be spontaneous or induced by environmental factors such as radiation, chemicals, or viruses, leading to alterations in genotype and phenotype.

Genetic Recombination: Genetic recombination occurs during meiosis, the process of cell division that gives rise to gametes (sperm and eggs). During recombination, homologous chromosomes exchange genetic material through crossing over, leading to the generation of novel combinations of alleles and increasing genetic diversity within populations.

Gene Flow: Gene flow refers to the exchange of genetic material between populations through migration and interbreeding. Gene flow can introduce new alleles into populations, homogenize genetic variation between populations, or facilitate adaptation to changing environments through the transfer of advantageous traits.

Genetic Drift: Genetic drift refers to the random fluctuations in allele frequencies within populations due to chance events, such as population bottlenecks or founder effects. Genetic drift can lead to the loss of genetic diversity and the fixation of alleles in small or isolated populations, shaping patterns of genetic variation over time.

Significance of Genetic Variation

Genetic variation serves as the raw material for evolution, enabling populations to adapt to changing environmental conditions, resist pathogens, and exploit new ecological niches. Understanding the types and sources of genetic variation is essential for addressing fundamental questions in genetics, evolutionary biology, and medicine, including:

Population Genetics: Genetic variation provides insights into the demographic history, migration patterns, and evolutionary dynamics of populations, shedding light on the processes that shape genetic diversity within and between species.

Human Genetics: Genetic variation underlies the diversity of human traits, susceptibility to diseases, and responses to pharmacological treatments. Studying genetic variation in human populations can inform personalized medicine approaches, disease risk assessment, and public health interventions.

Conservation Biology: Genetic variation plays a crucial role in the conservation and management of endangered species, guiding efforts to preserve genetic diversity, maintain population viability, and mitigate the effects of inbreeding and genetic drift.

Deploying Genetic Variation Analysis Techniques

To analyze genetic variation, researchers often utilize bioinformatics tools and software packages designed to process, annotate, and interpret genomic data. For instance, to identify SNPs from DNA sequencing data, researchers can use the GATK (Genome Analysis Toolkit) with the following CLI command:

bashCopy code

```
gatk HaplotypeCaller -R reference_genome.fasta -I input_reads.bam -O output_variants.vcf
```

This command runs the HaplotypeCaller tool from the GATK package on the input reads in BAM format, using the reference genome sequence provided in FASTA format, and outputs the detected variants in Variant Call Format (VCF).

Future Perspectives

As genomic technologies continue to advance and sequencing costs decline, the study of genetic variation is poised for unprecedented growth and innovation. By integrating multi-omics approaches, population-scale sequencing initiatives, and computational methods, researchers can gain deeper insights into the genetic basis of complex traits, disease susceptibility, and evolutionary processes. Leveraging genetic variation data holds immense promise for addressing global challenges in health, agriculture, and conservation, paving the way for a more sustainable and resilient future.

In summary, genetic variation represents the cornerstone of biological diversity and evolutionary change, shaping the genetic landscape of populations and species across the tree of life. By elucidating the types and sources of genetic variation, researchers can unravel the mysteries of genome evolution, adaptation, and inheritance, advancing our understanding of life's complexity and diversity.

SNP Analysis and Variation Detection Techniques

Single Nucleotide Polymorphisms (SNPs) are among the most prevalent forms of genetic variation found in the human genome and other organisms. Understanding SNPs and detecting their presence and distribution is crucial for numerous applications in genetics, from population studies to personalized medicine. This chapter explores SNP analysis techniques and variation detection methods, shedding light on their importance and deployment in genomic research.

Understanding SNPs

SNPs are variations at a single nucleotide position in the genome, where different nucleotides (A, T, C, or G) are present within a population. They are the most abundant type of genetic variation and can occur in both coding and non-coding regions of the genome. While many SNPs are neutral and have no discernible effect on phenotype, others can influence traits, disease susceptibility, and drug response.

SNP Analysis Techniques

Several techniques are employed for SNP analysis, each offering advantages and limitations depending on the research objectives, sample size, and available resources.

1. PCR-RFLP (Polymerase Chain Reaction-Restriction Fragment Length Polymorphism)

PCR-RFLP is a cost-effective method used to detect SNPs by amplifying a DNA fragment containing the SNP of interest using PCR, followed by digestion with a restriction enzyme that cleaves the DNA at specific recognition sites based on SNP alleles. The resulting fragments are then separated by gel electrophoresis to visualize the presence or absence of SNP alleles.

bashCopy code

```
PCR -i input_DNA_sequence.fasta -p primers.fasta -o amplified_fragment.fasta RestrictionEnzyme -i amplified_fragment.fasta -e enzyme_name -o digested_fragments.fasta
```

2. Allele-Specific PCR

Allele-specific PCR amplifies target DNA sequences using primers designed to specifically recognize SNP alleles. By designing primers that match the sequence of one allele but not the other, allele-specific PCR allows for the selective amplification of specific SNP alleles, enabling rapid and sensitive SNP detection.

bashCopy code

```
AS-PCR     -i     input_DNA_sequence.fasta     -p
allele_specific_primers.fasta -o amplified_product.fasta
```

3. SNP Microarrays

SNP microarrays, also known as SNP chips, are high-throughput platforms used to genotype thousands to millions of SNPs simultaneously. Microarrays consist of DNA probes complementary to specific SNP sequences, which are hybridized with target DNA samples. By measuring the intensity of fluorescent signals, SNP microarrays can identify SNP alleles present in the sample.

bashCopy code

```
SNP-Microarray -i input_DNA_sample.fasta -c chip_type
-o genotype_results.txt
```

4. Next-Generation Sequencing (NGS)

NGS technologies such as Illumina sequencing enable comprehensive SNP analysis by sequencing millions of DNA fragments in parallel. By aligning sequencing reads to a reference genome and identifying nucleotide variations, NGS platforms can detect SNPs with high accuracy and resolution, providing valuable insights into genetic diversity and population genetics.

bashCopy code

NGS-Alignment -i input_reads.fastq -r reference_genome.fasta -o aligned_reads.sam SNP-Calling -i aligned_reads.sam -o snp_calls.vcf

Significance of SNP Analysis

SNP analysis plays a crucial role in various fields of genetics and genomics, offering insights into population structure, evolutionary history, disease genetics, and personalized medicine.

Population Genetics: SNP analysis is used to study genetic variation within and between populations, revealing patterns of ancestry, migration, and adaptation. Population-level SNP data can inform conservation efforts, forensic genetics, and studies of human diversity.

Disease Genetics: SNPs are associated with numerous diseases and traits, serving as genetic markers for disease susceptibility, prognosis, and treatment response. Genome-wide association studies (GWAS) leverage SNP data to identify genetic variants associated with complex diseases such as cancer, diabetes, and cardiovascular disorders.

Pharmacogenomics: SNP analysis is used in pharmacogenomics to predict individual responses to drugs based on genetic variations. By identifying SNPs associated with drug metabolism, efficacy, and adverse reactions, personalized medicine approaches can optimize drug therapies and minimize side effects.

Evolutionary Biology: SNP analysis provides insights into the evolutionary history and adaptive strategies of organisms, elucidating the genetic basis of traits such as coat color, immune response, and environmental adaptation. Comparative SNP analysis across species can uncover signatures of natural selection and evolutionary divergence.

Challenges and Future Directions

Despite its utility, SNP analysis faces challenges such as data interpretation, allele frequency estimation, and distinguishing between causal variants and genetic linkage. Future advances in sequencing technologies, bioinformatics algorithms, and functional genomics will continue to enhance the accuracy, scalability, and applicability of SNP analysis techniques, unlocking new insights into the genetic basis of health and disease.

In summary, SNP analysis and variation detection techniques are indispensable tools in modern genetics, enabling researchers to explore genetic diversity, disease genetics, and evolutionary biology. By leveraging diverse SNP analysis methods and integrating multi-omics approaches, researchers can unravel the complexities of the genome and pave the way for personalized medicine and precision genomics.

Chapter 5: Introduction to Bioinformatics Algorithms

Algorithm Design and Analysis in Bioinformatics

Algorithm design and analysis form the backbone of bioinformatics, enabling researchers to develop efficient computational methods for solving complex biological problems. This chapter explores the principles of algorithm design and analysis in bioinformatics, showcasing their importance and practical application in genomic research and computational biology.

Understanding Algorithm Design

Algorithm design involves the creation of step-by-step procedures or instructions for solving a specific computational problem. In bioinformatics, algorithm design focuses on developing algorithms tailored to address challenges in analyzing biological data, such as DNA sequencing, sequence alignment, and protein structure prediction. Effective algorithm design requires a deep understanding of biological principles, mathematical concepts, and computational techniques.

Principles of Algorithm Design in Bioinformatics

Several key principles guide algorithm design in bioinformatics, including:

Efficiency: Bioinformatics algorithms must be efficient in terms of time and space complexity, capable of processing large-scale genomic datasets within reasonable computational resources. Efficiency is crucial for handling the vast amount of biological data generated by high-throughput sequencing technologies and other omics platforms.

Accuracy: Bioinformatics algorithms should produce accurate and reliable results, minimizing errors and false positives/negatives in data analysis. Accuracy is essential for interpreting genomic and proteomic data accurately, ensuring that computational predictions align with experimental observations.

Scalability: Bioinformatics algorithms should be scalable, able to accommodate increasing dataset sizes and computational demands as biological research progresses. Scalable algorithms can handle diverse applications, from genome assembly and annotation to phylogenetic analysis and drug discovery.

Robustness: Bioinformatics algorithms should be robust, capable of handling noisy or incomplete data, as well as accounting for biological variability and uncertainty. Robust algorithms can tolerate variations in experimental conditions, sample quality, and sequencing artifacts, ensuring robust performance across diverse datasets.

Practical Applications of Algorithm Design in Bioinformatics

Algorithm design finds numerous applications in bioinformatics, contributing to various aspects of genomic analysis, molecular modeling, and systems biology. Some practical applications of algorithm design in bioinformatics include:

Sequence Alignment: Sequence alignment algorithms, such as dynamic programming-based methods (e.g., Needleman-Wunsch, Smith-Waterman) and heuristic approaches (e.g., BLAST, Bowtie), are used to compare DNA, RNA, and protein sequences, identifying similarities, evolutionary relationships, and functional domains.

Genome Assembly: Genome assembly algorithms reconstruct complete genomes from fragmented DNA sequencing reads, leveraging graph-based approaches (e.g., de Bruijn graphs, overlap-layout-consensus) and statistical models (e.g., Eulerian paths, Bayesian inference) to assemble complex genomes accurately.

Protein Structure Prediction: Protein structure prediction algorithms use computational models and machine learning techniques to predict the three-dimensional structure of proteins from their amino acid sequences. Methods such as homology modeling, ab initio modeling, and molecular dynamics simulations are employed to predict protein folding, stability, and interactions.

Phylogenetic Analysis: Phylogenetic analysis algorithms reconstruct evolutionary trees and infer evolutionary relationships between species or genes, employing distance-based methods (e.g., neighbor-joining), character-based methods (e.g., maximum likelihood), and Bayesian inference to estimate phylogenetic trees from sequence data.

Deploying Algorithm Design Techniques

To deploy algorithm design techniques in bioinformatics, researchers utilize programming languages, software libraries, and bioinformatics tools tailored to specific algorithms and applications. For example:

Python: Python is a versatile programming language widely used in bioinformatics for algorithm implementation, data manipulation, and visualization. Libraries such as Biopython provide ready-to-use modules for sequence analysis, alignment, and phylogenetics.

R: R is a programming language and environment for statistical computing and data visualization, commonly used in bioinformatics for analyzing high-throughput sequencing data, performing statistical tests, and generating plots.

Bioinformatics Tools: Bioinformatics tools such as BLAST, ClustalW, and MUSCLE provide command-line interfaces and graphical user interfaces (GUIs) for running standard bioinformatics algorithms and analyses. For example:

bashCopy code

```
blastn -query query_sequence.fasta -subject subject_sequence.fasta -out alignment_results.txt
```

This command runs a BLAST search using the query sequence provided in "query_sequence.fasta" against the subject sequence in "subject_sequence.fasta" and outputs the alignment results to a text file named "alignment_results.txt".

Significance of Algorithm Design in Bioinformatics

Algorithm design plays a pivotal role in advancing our understanding of the biological world, enabling researchers to extract meaningful insights from genomic, transcriptomic, and proteomic data. By developing innovative algorithms and computational methods, bioinformaticians can unravel the complexities of biological systems, accelerate drug discovery, and improve human health.

Challenges and Future Directions

Despite its importance, algorithm design in bioinformatics faces challenges such as data integration, algorithm scalability, and computational efficiency. Future research directions in algorithm design may focus on developing machine learning algorithms, deep learning models, and network-based approaches to tackle complex biological problems and leverage large-scale omics datasets effectively.

In summary, algorithm design and analysis are essential components of bioinformatics, empowering researchers to develop computational solutions for analyzing biological data, understanding genetic mechanisms, and addressing biomedical challenges. By adhering to principles of efficiency, accuracy, scalability, and robustness, bioinformatics algorithms pave the way for groundbreaking discoveries in genetics, genomics, and systems biology.

Bioinformatics algorithms serve as the computational foundation for analyzing biological data, enabling researchers to extract meaningful insights from genomic, transcriptomic, and proteomic datasets. This chapter provides an overview of basic bioinformatics algorithms, illustrating their importance and practical application through examples and case studies.

Understanding Bioinformatics Algorithms

Bioinformatics algorithms encompass a wide range of computational techniques designed to solve specific biological problems, such as sequence alignment, genome assembly, gene prediction, and phylogenetic analysis. These algorithms leverage principles from computer science, mathematics, and statistics to process and interpret biological data, facilitating research in genetics, genomics, and systems biology.

Key Categories of Bioinformatics Algorithms

Bioinformatics algorithms can be categorized into several key areas, each addressing different aspects of biological data analysis:

Sequence Alignment Algorithms: Sequence alignment algorithms compare DNA, RNA, or protein sequences to identify similarities, evolutionary relationships, and functional motifs. Examples include dynamic programming algorithms (e.g., Needleman-Wunsch, Smith-Waterman) and heuristic methods (e.g., BLAST, Bowtie).

Genome Assembly Algorithms: Genome assembly algorithms reconstruct complete genomes from fragmented DNA sequencing reads, leveraging graph-based approaches (e.g., de Bruijn graphs, overlap-layout-consensus) and statistical models (e.g., Eulerian paths, Bayesian inference) to assemble complex genomes accurately.

Gene Prediction Algorithms: Gene prediction algorithms identify protein-coding genes and other functional elements within genomes, combining computational models (e.g., hidden Markov models, machine learning) with genomic features such as open reading frames, splice sites, and regulatory sequences.

Phylogenetic Analysis Algorithms: Phylogenetic analysis algorithms reconstruct evolutionary trees and infer evolutionary relationships between species or genes, employing distance-based methods (e.g., neighbor-joining), character-based methods (e.g., maximum likelihood), and Bayesian inference to estimate phylogenetic trees from sequence data.

Examples of Basic Bioinformatics Algorithms

Let's explore some examples of basic bioinformatics algorithms and their practical application:

1. Needleman-Wunsch Algorithm (Global Sequence Alignment)

The Needleman-Wunsch algorithm performs global sequence alignment, aligning two sequences to identify regions of similarity and dissimilarity. It is commonly used in comparative genomics, evolutionary analysis, and protein structure prediction.

bashCopy code

```
needleman_wunsch -sequence1 sequence1.fasta -sequence2 sequence2.fasta -output alignment_results.txt
```

This command runs the Needleman-Wunsch algorithm on the sequences provided in "sequence1.fasta" and "sequence2.fasta" and outputs the alignment results to a text file named "alignment_results.txt".

2. de Bruijn Graph Assembly Algorithm

The de Bruijn graph assembly algorithm is used for de novo genome assembly from DNA sequencing reads. It constructs a graph representation of overlapping k-mers from sequencing reads, followed by graph traversal to reconstruct the genome sequence.

bashCopy code

```
de_bruijn_assembly -reads input_reads.fastq -kmer_size 31 -output assembled_genome.fasta
```

This command performs de Bruijn graph assembly on the sequencing reads provided in "input_reads.fastq", using a k-mer size of 31, and outputs the assembled genome sequence to a FASTA file named "assembled_genome.fasta".

3. Hidden Markov Model (Gene Prediction)

Hidden Markov models (HMMs) are probabilistic models used for gene prediction in genomic sequences. They model the statistical properties of protein-coding genes, non-coding regions, and sequence features to predict gene structures accurately.

bashCopy code

```
hmm_gene_prediction -genome input_genome.fasta -hmm_model gene_model.hmm -output predicted_genes.gff
```

This command applies a hidden Markov model (stored in "gene_model.hmm") to predict genes in the genome sequence provided in "input_genome.fasta" and outputs the predicted gene structures in General Feature Format (GFF) to a file named "predicted_genes.gff".

4. Neighbor-Joining Algorithm (Phylogenetic Analysis)

The neighbor-joining algorithm constructs phylogenetic trees from distance matrices representing pairwise sequence similarities. It is widely used for inferring evolutionary relationships between species, populations, or genes based on sequence data.

bashCopy code

```
neighbor_joining                    -distance_matrix
input_distance_matrix.txt                   -output
phylogenetic_tree.nwk
```
This command applies the neighbor-joining algorithm to the distance matrix provided in "input_distance_matrix.txt" and outputs the inferred phylogenetic tree in Newick format to a file named "phylogenetic_tree.nwk".

Significance and Future Directions

Basic bioinformatics algorithms form the foundation of computational biology, driving discoveries in genetics, genomics, and evolutionary biology. As sequencing technologies continue to advance and genomic datasets grow in complexity, the development of novel algorithms and analytical methods will be critical for addressing emerging challenges in bioinformatics, such as analyzing single-cell omics data, integrating multi-omics datasets, and understanding the functional impact of genetic variation.

In summary, basic bioinformatics algorithms play a vital role in analyzing biological data, enabling researchers to decipher the genetic code, infer evolutionary relationships, and uncover the molecular mechanisms underlying life's diversity. By mastering fundamental algorithmic techniques and leveraging computational tools, bioinformaticians can advance our understanding of the biological world and drive innovation in biomedicine, agriculture, and environmental science.

Chapter 6: Pairwise Sequence Alignment Techniques

The Needleman-Wunsch algorithm stands as a cornerstone in bioinformatics, offering a powerful method for sequence alignment, a fundamental task in genomic analysis. Next, we delve into the intricacies of the Needleman-Wunsch algorithm, exploring its principles, applications, and practical deployment through command-line examples and explanations.

Understanding the Needleman-Wunsch Algorithm

The Needleman-Wunsch algorithm is a dynamic programming-based approach used for global sequence alignment. Its primary objective is to align two sequences, typically DNA, RNA, or protein sequences, to identify regions of similarity and dissimilarity. By assigning scores to match, mismatch, and gap penalties, the algorithm computes the optimal alignment between the sequences, maximizing their similarity while minimizing the number of gaps introduced.

Principles of the Needleman-Wunsch Algorithm

The Needleman-Wunsch algorithm operates on a dynamic programming matrix, where each cell represents the optimal alignment score between corresponding segments of the two sequences. By iteratively filling the matrix and tracing back the optimal alignment path, the algorithm identifies the optimal alignment and computes its alignment score.

Practical Application of the Needleman-Wunsch Algorithm

Let's explore a practical application of the Needleman-Wunsch algorithm for aligning two DNA sequences using command-line tools:

bashCopy code

needleman_wunsch -sequence1 sequence1.fasta -sequence2 sequence2.fasta -output alignment_results.txt

In this command, "needleman_wunsch" invokes the Needleman-Wunsch algorithm, aligning the sequences provided in "sequence1.fasta" and "sequence2.fasta". The alignment results are then outputted to a text file named "alignment_results.txt".

Example Scenario: DNA Sequence Alignment

Suppose we have two DNA sequences:

Sequence 1: AGTACGCTAG Sequence 2: ACTAGCTGA

Using the Needleman-Wunsch algorithm with a scoring scheme of +1 for matches, -1 for mismatches, and -2 for gap penalties, let's compute the optimal alignment:

mathematicaCopy code

Sequence 1: AGTACGCTAG Sequence 2: A-CTAGCTGA

Alignment Score: +5

In this alignment, the optimal alignment score is +5, achieved by aligning the sequences as shown above. The algorithm identifies the best alignment by considering all possible alignment paths and selecting the one with the highest score.

Significance of the Needleman-Wunsch Algorithm

The Needleman-Wunsch algorithm holds immense significance in bioinformatics and computational biology:

Sequence Comparison: The algorithm enables researchers to compare DNA, RNA, or protein sequences, facilitating the identification of evolutionary relationships, conserved motifs, and functional domains.

Genome Annotation: In genome annotation, the Needleman-Wunsch algorithm is used to align gene sequences against reference genomes, aiding in the identification of gene structures, regulatory elements, and genetic variations.

Protein Structure Prediction: The algorithm plays a crucial role in protein structure prediction, aligning protein sequences to infer their three-dimensional structures and functional properties.

Evolutionary Analysis: By aligning sequences from different species or homologous genes, the Needleman-Wunsch algorithm helps elucidate evolutionary histories, genetic divergence, and adaptive evolution.

Future Directions and Challenges

While the Needleman-Wunsch algorithm has revolutionized sequence alignment, challenges remain in optimizing its computational efficiency and scalability for analyzing large-scale genomic datasets. Future research directions may focus on developing parallelized algorithms, heuristic approaches, and machine learning techniques to accelerate sequence alignment and address emerging challenges in bioinformatics.

In summary, the Needleman-Wunsch algorithm stands as a fundamental tool in bioinformatics, empowering researchers to unravel the mysteries of the genome and decode the genetic basis of life. By mastering the principles of dynamic programming and sequence alignment, bioinformaticians can advance our understanding of biological systems, drive innovation in biomedicine, and tackle pressing challenges in health, agriculture, and environmental science.

Smith-Waterman Algorithm: Local Alignment
The Smith-Waterman algorithm represents a pivotal advancement in bioinformatics, offering a robust method for local sequence alignment, a crucial task in genomic analysis. This chapter delves into the intricacies of the Smith-Waterman algorithm, elucidating its principles, applications, and practical deployment through command-line examples and detailed explanations.

Understanding the Smith-Waterman Algorithm
The Smith-Waterman algorithm serves as a cornerstone in bioinformatics for local sequence alignment. Unlike the Needleman-Wunsch algorithm, which focuses on global alignment, the Smith-Waterman algorithm identifies local regions of similarity between two sequences, allowing for the detection of conserved motifs, domains, and functional elements.

Principles of the Smith-Waterman Algorithm
The Smith-Waterman algorithm operates on a dynamic programming matrix similar to the Needleman-Wunsch algorithm. However, it introduces an additional step called traceback, enabling the identification of the optimal local alignment by tracing back the highest-

scoring submatrix within the dynamic programming matrix.

Practical Application of the Smith-Waterman Algorithm

Let's explore a practical application of the Smith-Waterman algorithm for aligning two protein sequences using command-line tools:

bashCopy code

```
smith_waterman -sequence1 protein1.fasta -sequence2 protein2.fasta -output local_alignment.txt
```

In this command, "smith_waterman" invokes the Smith-Waterman algorithm, aligning the protein sequences provided in "protein1.fasta" and "protein2.fasta". The local alignment results are then outputted to a text file named "local_alignment.txt".

Example Scenario: Protein Sequence Alignment

Suppose we have two protein sequences representing a protein domain:

Sequence 1: MAEEAVLWKGFGG Sequence 2: LWWKGF

Using the Smith-Waterman algorithm with a scoring scheme of +1 for matches, -1 for mismatches, and -2 for gap penalties, let's compute the optimal local alignment:

mathematicaCopy code

```
Sequence 1: MAEEAVLWKGFGG Sequence 2: LW-KGF
Alignment Score: +4
```

In this alignment, the optimal local alignment score is +4, achieved by aligning the segments "LWKG" in both sequences. The Smith-Waterman algorithm identifies this local alignment as the highest-scoring subsequence within the sequences.

Significance of the Smith-Waterman Algorithm

The Smith-Waterman algorithm holds profound significance in bioinformatics and computational biology:

Motif and Domain Identification: The algorithm enables researchers to identify conserved motifs, domains, and functional elements within biological sequences, shedding light on protein structure, function, and evolution.

Sequence Database Search: In sequence database search applications, such as protein database searches, the Smith-Waterman algorithm is used to identify local similarities between query sequences and database entries, facilitating the annotation of unknown proteins and genes.

Structural Bioinformatics: The algorithm plays a crucial role in structural bioinformatics, aligning protein sequences to predict their three-dimensional structures and infer functional relationships between proteins.

Variant Detection: In genomic analysis, the Smith-Waterman algorithm can be used to identify local sequence variations, such as single nucleotide polymorphisms (SNPs) and insertions/deletions (indels), by comparing sequences from different individuals or species.

Future Directions and Challenges

While the Smith-Waterman algorithm has revolutionized local sequence alignment, challenges remain in optimizing its computational efficiency and scalability for analyzing large-scale genomic datasets. Future research directions may focus on developing parallelized algorithms, heuristic approaches, and hardware-accelerated implementations to accelerate local sequence alignment and address emerging challenges in bioinformatics.

In summary, the Smith-Waterman algorithm stands as a cornerstone in bioinformatics, empowering researchers to explore the intricate details of biological sequences and unravel the mysteries of the genome. By mastering the principles of dynamic programming and local sequence alignment, bioinformaticians can advance our understanding of protein structure, function, and evolution, driving innovation in biomedicine, agriculture, and environmental science.

Chapter 7: Multiple Sequence Alignment Methods

ClustalW and Clustal Omega represent two prominent tools in bioinformatics for multiple sequence alignment, a fundamental task in genomic analysis. This chapter delves into the intricacies of ClustalW and Clustal Omega, elucidating their principles, applications, and practical deployment through command-line examples and detailed explanations.

Understanding ClustalW and Clustal Omega

ClustalW and Clustal Omega are software packages developed for multiple sequence alignment, a process that aligns three or more biological sequences (DNA, RNA, or protein) to identify regions of similarity and divergence. By aligning sequences, researchers can infer evolutionary relationships, identify conserved motifs, and predict functional domains within biological molecules.

Principles of ClustalW and Clustal Omega

Both ClustalW and Clustal Omega operate on progressive alignment algorithms, which construct a guide tree to sequentially align sequences based on their pairwise similarities. While ClustalW utilizes a slower and less accurate algorithm, Clustal Omega employs a more efficient and accurate approach, making it suitable for aligning large-scale datasets.

Practical Application of ClustalW and Clustal Omega

Let's explore a practical application of ClustalW and Clustal Omega for aligning protein sequences using command-line tools:

bashCopy code

```
clustalw -infile sequences.fasta -outfile alignment.clw
```

In this command, "clustalw" invokes the ClustalW software, aligning the protein sequences provided in "sequences.fasta". The alignment results are then outputted to a file named "alignment.clw".

Similarly, for Clustal Omega:

bashCopy code

```
clustalo -i sequences.fasta -o alignment.fasta --outfmt=fasta
```

In this command, "clustalo" invokes Clustal Omega, aligning the protein sequences provided in "sequences.fasta". The alignment results are outputted to a file named "alignment.fasta" in FASTA format.

Example Scenario: Protein Sequence Alignment

Suppose we have three protein sequences representing homologous genes:

Sequence 1: MKEIKNWEFQ Sequence 2: MKEIGKWTFQ Sequence 3: MKEVGNWDFK

Using ClustalW and Clustal Omega, let's compute the multiple sequence alignment for these sequences:

objectivecCopy code

```
ClustalW:        MKEIKNWEFQ        MKEIGKWTFQ
MKEVGNWDFK   Clustal   Omega:   MKEIKNWEFQ
MKEIGKWTFQ  MKEVGNWDFK
```

Both ClustalW and Clustal Omega align the sequences, identifying conserved residues and insertions/deletions across the sequences. While Clustal Omega may produce a more accurate alignment due to its improved algorithm, both tools provide valuable insights into the evolutionary relationships between the sequences.

Significance of ClustalW and Clustal Omega

ClustalW and Clustal Omega play a pivotal role in bioinformatics and computational biology:

Phylogenetic Analysis: The alignments generated by ClustalW and Clustal Omega serve as inputs for phylogenetic analysis, enabling researchers to reconstruct evolutionary trees and infer genetic relationships between species or genes.

Structural Bioinformatics: Multiple sequence alignments are essential for structural bioinformatics, aiding in the prediction of protein structures, identification of functional domains, and analysis of protein-protein interactions.

Functional Annotation: By identifying conserved residues and motifs, ClustalW and Clustal Omega facilitate the functional annotation of genes and proteins, guiding experimental studies on protein function and biological pathways.

Variant Analysis: In genomic analysis, multiple sequence alignments are used to compare sequences from different individuals or species, facilitating the identification of genetic variations, such as single nucleotide polymorphisms (SNPs) and indels.

Future Directions and Challenges

While ClustalW and Clustal Omega represent state-of-the-art tools for multiple sequence alignment, challenges remain in optimizing their algorithms for analyzing large-scale genomic datasets and improving their accuracy for aligning distantly related sequences.

Future research directions may focus on developing hybrid alignment methods, integrating machine learning techniques, and leveraging cloud computing resources to address these challenges and advance the field of sequence analysis.

In summary, ClustalW and Clustal Omega stand as indispensable tools in bioinformatics, empowering researchers to analyze and interpret biological sequences with unprecedented accuracy and efficiency. By mastering the principles of multiple sequence alignment and leveraging advanced computational tools, bioinformaticians can unravel the mysteries of the genome, decipher the genetic code of life, and drive innovation in biomedicine, agriculture, and environmental science.

Progressive and Iterative Alignment Strategies: Unraveling Complex Sequence Relationships

Progressive and iterative alignment strategies represent two fundamental approaches in bioinformatics for aligning multiple sequences, a critical task in genomic analysis. This chapter delves into the intricacies of progressive and iterative alignment strategies, elucidating their principles, applications, and practical deployment through command-line examples and detailed explanations.

Understanding Progressive and Iterative Alignment Strategies

Progressive alignment strategies, such as those implemented in ClustalW, construct a guide tree to sequentially align sequences based on their pairwise similarities. In contrast, iterative alignment strategies, exemplified by algorithms like MAFFT and T-Coffee, refine alignments through iterative refinement steps, incorporating information from previous alignments to improve accuracy and consistency.

Principles of Progressive Alignment

Progressive alignment begins with pairwise sequence comparisons to calculate a distance matrix, which is then used to construct a guide tree representing the evolutionary relationships between sequences. The sequences are then aligned sequentially along the branches of the guide tree, starting from the most closely related pairs and gradually incorporating more distant sequences into the alignment.

Principles of Iterative Alignment

Iterative alignment strategies iteratively refine initial alignments using iterative refinement steps. These steps typically involve aligning sequences, calculating a consensus alignment, and using the consensus alignment as a reference for subsequent iterations. By iteratively refining alignments, iterative strategies can improve alignment accuracy and consistency, particularly for distantly related sequences.

Practical Application of Progressive and Iterative Alignment Strategies

Let's explore a practical application of progressive and iterative alignment strategies for aligning protein sequences using command-line tools:

Progressive Alignment (ClustalW):

bashCopy code

```
clustalw    -infile    sequences.fasta    -outfile
progressive_alignment.clw
```

In this command, "clustalw" invokes the ClustalW software, performing progressive alignment on the protein sequences provided in "sequences.fasta". The progressive alignment results are then outputted to a file named "progressive_alignment.clw".

Iterative Alignment (MAFFT):

bashCopy code

```
mafft    --auto    sequences.fasta    >
iterative_alignment.fasta
```

In this command, "mafft" invokes the MAFFT software, performing iterative alignment on the protein sequences provided in "sequences.fasta". The iterative alignment results are then outputted to a file named "iterative_alignment.fasta".

Example Scenario: Protein Sequence Alignment

Suppose we have four protein sequences representing homologous genes:

Sequence 1: MKEIKNWEFQ Sequence 2: MKEIGKWTFQ Sequence 3: MKEVGNWDFK Sequence 4: MKEAGNWDFR

Using both progressive and iterative alignment strategies, let's compute the multiple sequence alignment for these sequences:

objectivecCopy code

Progressive Alignment: MKEIKNWEFQ MKEIGKWTFQ MKEVGNWDFK MKEAGNWDFR Iterative Alignment: MKEIKNWEFQ   MKEIGKWTFQ   MKEVGNWDFK MKEAGNWDFR

Both progressive and iterative alignment strategies align the sequences, identifying conserved residues and insertions/deletions across the sequences. However, iterative alignment strategies may produce more accurate alignments, particularly for distantly related sequences.

Significance of Progressive and Iterative Alignment Strategies

Progressive and iterative alignment strategies play a crucial role in bioinformatics and computational biology:

Phylogenetic Analysis: The alignments generated by progressive and iterative strategies serve as inputs for phylogenetic analysis, enabling researchers to reconstruct evolutionary trees and infer genetic relationships between species or genes.

Structural Bioinformatics: Multiple sequence alignments are essential for structural bioinformatics, aiding in the prediction of protein structures, identification of functional domains, and analysis of protein-protein interactions.

Functional Annotation: By identifying conserved residues and motifs, progressive and iterative alignment strategies facilitate the functional annotation of genes and proteins, guiding experimental studies on protein function and biological pathways.

Variant Analysis: In genomic analysis, multiple sequence alignments are used to compare sequences from different individuals or species, facilitating the identification of genetic variations, such as single nucleotide polymorphisms (SNPs) and indels.

Future Directions and Challenges

While progressive and iterative alignment strategies have revolutionized multiple sequence alignment, challenges remain in optimizing their algorithms for analyzing large-scale genomic datasets and improving their accuracy for aligning distantly related sequences. Future research directions may focus on developing hybrid alignment methods, integrating machine learning techniques, and leveraging cloud computing resources to address these challenges and advance the field of sequence analysis.

In summary, progressive and iterative alignment strategies stand as indispensable tools in bioinformatics, empowering researchers to analyze and interpret biological sequences with unprecedented accuracy and efficiency. By mastering the principles of multiple sequence alignment and leveraging advanced computational tools, bioinformaticians can unravel the mysteries of the genome, decipher the genetic code of life, and drive innovation in biomedicine, agriculture, and environmental science.

Chapter 8: Genome Assembly and Annotation

Genome assembly is a critical process in bioinformatics that involves reconstructing the complete genome sequence of an organism from fragmented DNA sequencing reads. This chapter explores the principles, applications, and deployment of two primary genome assembly approaches: de novo assembly and reference-based assembly. Through detailed explanations and command-line examples, readers will gain insights into the strengths, limitations, and practical considerations of each approach.

Understanding Genome Assembly

Genome assembly aims to reconstruct the original DNA sequence of an organism by piecing together overlapping DNA fragments obtained from high-throughput sequencing technologies. This process is essential for studying genomic structure, identifying genetic variations, and understanding the functional elements encoded within the genome.

De Novo Assembly

De novo assembly is a genome assembly approach that reconstructs the genome sequence without relying on a reference genome. The process begins by assembling overlapping sequencing reads into contiguous sequences called contigs, using algorithms such as de Bruijn graph-based methods or overlap-layout-consensus approaches. These contigs are then scaffolded and oriented to generate longer sequences representing chromosomal segments. De novo assembly is particularly useful for non-model organisms or species with complex or highly variable genomes where a reference genome is not available.

Practical Deployment of De Novo Assembly

Let's consider an example of de novo assembly using the SPAdes software:

bashCopy code

```
spades.py       -1       forward_reads.fastq       -2
reverse_reads.fastq -o de_novo_assembly_output
```

In this command, "spades.py" invokes the SPAdes assembler, which performs de novo assembly using paired-end sequencing reads provided in "forward_reads.fastq" and "reverse_reads.fastq". The output of the assembly process is stored in the directory "de_novo_assembly_output".

Reference-Based Assembly

Reference-based assembly is an approach that aligns sequencing reads to a known reference genome to reconstruct the genome sequence. This method leverages the availability of a closely related reference genome to map sequencing reads, identify genetic variations, and fill gaps in the assembly. Reference-based assembly is efficient and accurate, particularly for well-studied organisms with high-quality reference genomes.

Practical Deployment of Reference-Based Assembly

Let's consider an example of reference-based assembly using the BWA-MEM software:

bashCopy code

```
bwa index reference_genome.fasta bwa mem
reference_genome.fasta input_reads.fastq >
alignment.sam samtools view -bS alignment.sam >
alignment.bam samtools sort alignment.bam -o
alignment_sorted.bam samtools index
alignment_sorted.bam
```

In these commands, we first index the reference genome using "bwa index". Then, we align the sequencing reads to the reference genome using "bwa mem", producing a Sequence Alignment/Map (SAM) file. We convert the SAM file to a Binary Alignment/Map (BAM) file using "samtools view", sort the BAM file using "samtools sort", and index the sorted BAM file using "samtools index".

Comparing De Novo and Reference-Based Assembly

De novo assembly offers the advantage of reconstructing genomes without the need for a reference genome, making it suitable for novel or understudied organisms. However, de novo assembly can be challenging for complex genomes or repetitive regions due to the high computational and memory requirements. In contrast, reference-based assembly provides a rapid and accurate reconstruction of genomes, but it relies on the availability of a closely related reference genome, limiting its applicability to well-characterized organisms.

In summary, genome assembly is a crucial step in genomic analysis, enabling researchers to reconstruct the complete DNA sequence of an organism. De novo assembly and reference-based assembly represent two primary approaches for genome assembly, each offering distinct advantages and challenges. By understanding the principles and practical considerations of these assembly approaches, bioinformaticians can effectively reconstruct genomes and unlock the secrets encoded within the genetic code.

Genome Annotation: Unlocking the Genetic Blueprint
Genome annotation is a crucial process in bioinformatics that involves identifying and annotating the functional elements encoded within a genome sequence. This chapter explores the principles, techniques, and practical deployment of genome annotation, focusing on functional annotation and gene prediction. Through detailed explanations and command-line examples, readers will gain insights into the tools, strategies, and challenges associated with genome annotation.

Understanding Genome Annotation
Genome annotation is the process of deciphering the biological significance of the nucleotide sequences within a genome. It involves identifying genes, regulatory elements, non-coding RNAs, and other functional elements that contribute to the organism's phenotype. Genome annotation provides valuable insights into gene function, evolutionary relationships, and the underlying genetic mechanisms driving biological processes.

Functional Annotation

Functional annotation involves assigning biological functions to the genes and other genomic elements identified within a genome sequence. This process relies on various bioinformatics tools and databases to predict the functions of genes based on sequence similarity, protein domain analysis, and other bioinformatic techniques. Functional annotation provides critical information about the roles and interactions of genes in cellular processes, pathways, and disease mechanisms.

Practical Deployment of Functional Annotation

Let's consider an example of functional annotation using the NCBI BLAST software:

bashCopy code

```
blastp -query protein_sequence.fasta -db nr -out blast_results.txt -evalue 0.001 -max_target_seqs 10 -outfmt 6
```

In this command, "blastp" performs a protein-protein BLAST search using a query protein sequence provided in "protein_sequence.fasta" against the NCBI non-redundant protein database (nr). The output of the BLAST search is stored in the file "blast_results.txt" in tabular format, with additional parameters specifying the maximum E-value threshold and the maximum number of target sequences to report.

Gene Prediction

Gene prediction is the process of identifying the protein-coding genes within a genome sequence. This task is challenging due to the presence of non-coding regions, alternative splicing, and overlapping genes. Gene prediction algorithms use statistical models, sequence motifs, and comparative genomics approaches to identify potential coding regions and predict the gene structures, including exon-intron boundaries and translation start and stop sites.

Practical Deployment of Gene Prediction

Let's consider an example of gene prediction using the Augustus software:

bashCopy code

augustus --species=human genome_sequence.fasta

In this command, "augustus" predicts genes within a genome sequence provided in "genome_sequence.fasta" using a species-specific model for humans. The output of the gene prediction process includes predicted gene structures, including exon-intron boundaries, coding sequences, and protein translations.

Comparing Functional Annotation and Gene Prediction

Functional annotation and gene prediction are complementary processes that provide valuable insights into the biological significance of genomic sequences. Functional annotation focuses on assigning biological functions to known genes and genomic elements, while gene prediction aims to identify novel protein-coding genes and predict their structures. Together, these processes enable researchers to unravel the complex genetic architecture of organisms and understand the molecular basis of biological processes and disease.

In summary, genome annotation is a critical step in genomic analysis, providing essential information about the genetic blueprint of organisms. Functional annotation and gene prediction are key components of genome annotation, enabling researchers to decipher the biological significance of genomic sequences and identify the genes and regulatory elements that underlie cellular processes. By mastering the principles and techniques of genome annotation, bioinformaticians can advance our understanding of genomics, biology, and human health.

Chapter 9: Phylogenetic Analysis

Phylogenetic trees are powerful tools in bioinformatics used to visualize and interpret the evolutionary relationships between biological entities, such as species, genes, or proteins. This chapter delves into the principles, methods, and practical deployment of phylogenetic tree construction and interpretation, offering insights into the evolutionary history of life on Earth.

Understanding Phylogenetic Trees

Phylogenetic trees depict the evolutionary relationships among organisms based on shared ancestry and genetic similarity. These trees are hierarchical structures, with branches representing evolutionary lineages and nodes indicating common ancestors. Phylogenetic trees provide valuable insights into the diversification, adaptation, and evolutionary history of species and genes.

Construction of Phylogenetic Trees

Phylogenetic trees are constructed using various methods, including distance-based methods, maximum parsimony, and maximum likelihood. Distance-based methods measure the genetic distances between sequences and construct trees based on these distances. Maximum parsimony aims to find the tree that requires the fewest evolutionary changes to explain the observed data. Maximum likelihood estimates the probability of different evolutionary models and chooses the tree that best fits the data.

Practical Deployment of Phylogenetic Tree Construction

Let's consider an example of phylogenetic tree construction using the Neighbor-Joining algorithm implemented in the PHYLIP software:

bashCopy code

```
neighbor < alignment.phy > tree_output.newick
```

In this command, "neighbor" executes the Neighbor-Joining algorithm using a multiple sequence alignment provided in "alignment.phy". The output of the algorithm is stored in the Newick format in the file "tree_output.newick", representing the phylogenetic tree.

Interpretation of Phylogenetic Trees

Phylogenetic trees provide valuable information about evolutionary relationships, divergence times, and ancestral traits. The branching pattern of the tree indicates the relatedness of species or genes, with closely related entities sharing a common ancestor and forming clusters or clades. Branch lengths can represent genetic distances or evolutionary time, with longer branches indicating greater divergence.

Practical Deployment of Phylogenetic Tree Interpretation

Let's consider an example of phylogenetic tree interpretation using the ETE Toolkit in Python:

pythonCopy code

```
from ete3 import Tree tree = Tree("tree_output.newick") print(tree)
```

In this Python script, the ETE Toolkit is used to read the Newick-formatted phylogenetic tree from the file "tree_output.newick" and display it in a graphical format. The resulting tree can be further analyzed to identify clades, measure branch lengths, and infer evolutionary relationships.

Applications of Phylogenetic Trees

Phylogenetic trees have diverse applications in biology, including:

Evolutionary Biology: Phylogenetic trees are used to study the evolutionary history of species, infer ancestral traits, and trace the origins of biological diversity.

Taxonomy and Classification: Phylogenetic trees provide a framework for taxonomic classification, helping to define species boundaries and organize biodiversity.

Comparative Genomics: Phylogenetic trees are used to compare the genomes of different organisms, identify orthologous genes, and infer gene function and evolution.

Biomedical Research: Phylogenetic trees are used to study the evolution of pathogens, trace the spread of infectious diseases, and develop strategies for disease prevention and treatment.

Future Directions and Challenges

While phylogenetic trees are powerful tools for studying evolutionary relationships, challenges remain in tree reconstruction, model selection, and interpretation. Future research directions may focus on developing more accurate and efficient algorithms, integrating genomic data with other types of biological data, and exploring new visualization and analysis techniques.

In summary, phylogenetic trees are invaluable tools for understanding the evolutionary history of life on Earth. By reconstructing and interpreting phylogenetic trees, researchers can unravel the complexities of evolutionary relationships, gain insights into the origins of biological diversity, and inform conservation efforts, biotechnology applications, and biomedical research.

Distance-Based and Character-Based Methods in Phylogenetics

In the realm of phylogenetics, the study of evolutionary relationships among organisms, two fundamental approaches emerge: distance-based and character-based methods. This chapter explores the principles, techniques, and practical applications of these methods, shedding light on their strengths, limitations, and relevance in modern bioinformatics.

Understanding Distance-Based Methods

Distance-based methods infer phylogenetic trees by calculating genetic distances between sequences and constructing trees based on these distances. These methods are conceptually straightforward and computationally efficient, making them suitable for large datasets. Common distance metrics include Hamming distance for nucleotide sequences and Jukes-Cantor or Kimura models for estimating genetic distances. Neighbor-Joining and UPGMA (Unweighted Pair Group Method with Arithmetic Mean) are popular algorithms for tree construction based on distance matrices.

Practical Deployment of Distance-Based Methods

Let's consider an example of deploying the Neighbor-Joining algorithm using the PHYLIP software:

bashCopy code

```
neighbor < alignment.phy > tree_output.newick
```

In this command, "neighbor" executes the Neighbor-Joining algorithm using a multiple sequence alignment provided in "alignment.phy". The resulting phylogenetic tree is stored in the Newick format in the file "tree_output.newick".

Understanding Character-Based Methods

Character-based methods, also known as parsimony methods, infer phylogenetic trees by minimizing the number of evolutionary changes required to explain observed sequence data. These methods analyze the presence or absence of characters, such as nucleotides or amino acids, across sequences and search for the most parsimonious tree that explains the observed character distribution. Maximum Parsimony and Maximum Likelihood are common character-based methods used for phylogenetic inference.

Practical Deployment of Character-Based Methods

Let's consider an example of deploying the Maximum Parsimony method using the PAUP* software:

bashCopy code

```
paup -n < input_file.nex > tree_output.nex
```

In this command, "paup" executes the Maximum Parsimony analysis using a NEXUS format input file containing sequence data and associated character states. The resulting phylogenetic tree is stored in the NEXUS format in the file "tree_output.nex".

Comparing Distance-Based and Character-Based Methods

Distance-based methods are advantageous for their computational efficiency and ability to handle large datasets, making them suitable for preliminary analyses and exploratory studies. However, they may oversimplify evolutionary processes and ignore phylogenetic signal present in character data. Character-based methods, on the other hand, explicitly model evolutionary changes and are robust to certain types of data heterogeneity, but they can be computationally intensive and sensitive to model assumptions.

Applications of Distance-Based and Character-Based Methods

Both distance-based and character-based methods find applications across various domains of biology, including:

Phylogenomics: Inferring evolutionary relationships among genes and genomes to study genome evolution and gene function.

Systematics: Reconstructing evolutionary trees to classify and organize biodiversity, aiding in taxonomic classification and species identification.

Molecular Epidemiology: Tracking the spread of infectious diseases by analyzing the evolutionary history and transmission patterns of pathogens.

Evolutionary Ecology: Investigating the evolutionary processes driving adaptation, speciation, and diversification in natural populations.

Future Directions and Challenges

Advancements in computational algorithms, statistical methods, and data integration techniques are driving innovation in phylogenetics. Future research directions may focus on developing hybrid methods that combine the strengths of distance-based and character-based approaches, integrating genomic data with other types of biological data, and exploring novel methods for modeling complex evolutionary processes.

In summary, distance-based and character-based methods are foundational tools in phylogenetics, enabling researchers to reconstruct evolutionary trees and unravel the complex history of life on Earth. By understanding the principles and practical applications of these methods, bioinformaticians can gain insights into the evolutionary relationships among organisms, elucidate the processes driving biological diversity, and address key questions in evolutionary biology and ecology.

Chapter 10: Applications of Bioinformatics in Biomedical Researc

In the landscape of biomedical research, disease gene identification and drug discovery stand as pivotal endeavors aimed at elucidating the genetic basis of diseases and developing effective therapeutic interventions. This chapter delves into the methodologies, strategies, and practical applications involved in identifying disease genes and discovering potential drug targets, offering insights into the intersection of genomics and therapeutics.

Understanding Disease Gene Identification

Disease gene identification involves pinpointing genetic variants or mutations associated with human diseases. This process often begins with genome-wide association studies (GWAS) or linkage analyses to identify genomic regions linked to disease susceptibility. Next-generation sequencing technologies, such as whole-exome sequencing (WES) and whole-genome sequencing (WGS), enable researchers to identify causal variants within these regions. Bioinformatics tools and databases, such as ANNOVAR and ClinVar, facilitate the annotation and interpretation of genetic variants, aiding in the identification of disease-causing mutations.

Practical Deployment of Disease Gene Identification

Let's consider an example of disease gene identification using whole-exome sequencing data and the ANNOVAR software:

bashCopy code

```
annotate_variation.pl   -downdb   -buildver   hg19   -
webfrom       annovar       refGene       humandb/
annotate_variation.pl -buildver hg19 -dbtype avsnp147
input.vcf   humandb/   table_annovar.pl   input.vcf
humandb/  -buildver  hg19  -out  output  -remove  -
protocol refGene,avsnp147 -operation g,f
```

In these commands, "annotate_variation.pl" downloads and installs annotation databases for human genome assembly version hg19, including gene annotations and single nucleotide polymorphism (SNP) annotations from ANNOVAR's web server. The "table_annovar.pl" command annotates variant call format (VCF) files with gene annotations and SNP annotations, producing an annotated output file.

Understanding Drug Discovery

Drug discovery involves identifying and developing compounds that can modulate biological targets associated with diseases. This process encompasses target identification, lead compound identification, preclinical testing, and clinical trials. Target identification often involves identifying disease-associated genes or proteins that play key roles in disease pathogenesis. High-throughput screening (HTS) and computational modeling techniques, such as molecular docking and virtual screening, are used to identify small molecules or biologics that interact with these targets and modulate their activity.

Practical Deployment of Drug Discovery Techniques

Let's consider an example of virtual screening for drug discovery using the Autodock Vina software:

```bash
bashCopy code
vina --config config.txt --ligand ligand.pdbqt --receptor
receptor.pdbqt --out output.pdbqt
```
In this command, "vina" performs virtual screening by docking the ligand molecule (ligand.pdbqt) to the receptor protein structure (receptor.pdbqt) using parameters specified in the configuration file (config.txt). The output file (output.pdbqt) contains the predicted binding poses and binding affinities of the ligand molecules.

Applications of Disease Gene Identification and Drug Discovery

Disease gene identification and drug discovery have transformative implications for human health and medicine, including:

Precision Medicine: Identifying disease-associated genetic variants enables the development of personalized therapies tailored to individual patients' genetic profiles.

Drug Repurposing: Understanding the genetic basis of diseases facilitates the repurposing of existing drugs for new indications, accelerating the drug development process.

Targeted Therapies: Identifying disease-associated genes and proteins enables the development of targeted therapies that selectively modulate disease pathways, minimizing off-target effects and improving therapeutic outcomes.

Pharmacogenomics: Understanding how genetic variations influence drug response enables the optimization of drug dosing and selection based on patients' genetic profiles, leading to safer and more effective treatments.

Challenges and Future Directions

Despite significant advancements in disease gene identification and drug discovery, several challenges remain, including the need for improved computational methods for variant interpretation, the development of more efficient drug screening technologies, and the translation of basic research findings into clinically actionable therapies. Future research directions may focus on leveraging big data analytics, artificial intelligence, and high-throughput experimental technologies to accelerate the pace of discovery and innovation in precision medicine.

In summary, disease gene identification and drug discovery represent two interconnected pillars of biomedical research, offering promise for advancing our understanding of disease mechanisms and developing targeted therapies for a wide range of human diseases. By integrating genomics, bioinformatics, and drug development approaches, researchers can unlock new insights into disease pathogenesis, identify novel therapeutic targets, and ultimately improve patient outcomes in the era of precision medicine.

Personalized Medicine: Transforming Healthcare Through Genomics

Personalized medicine, also known as precision medicine, has emerged as a revolutionary approach to healthcare, aiming to tailor medical treatment and prevention strategies to individual patients based on their unique genetic makeup, lifestyle, and environmental factors. This chapter explores the principles, methodologies, and clinical applications of personalized medicine, shedding light on its transformative potential in improving patient outcomes and revolutionizing healthcare delivery.

Understanding Personalized Medicine

At its core, personalized medicine recognizes that each patient is unique, with distinct genetic variations, biological pathways, and responses to treatments. By leveraging advances in genomics, bioinformatics, and data analytics, personalized medicine seeks to elucidate the genetic basis of diseases, identify biomarkers of disease susceptibility and drug response, and optimize treatment strategies to maximize efficacy and minimize adverse effects.

Genomic Profiling and Biomarker Discovery

Genomic profiling lies at the heart of personalized medicine, enabling the identification of genetic variations associated with disease susceptibility, prognosis, and treatment response. High-throughput sequencing technologies, such as whole-genome sequencing (WGS) and whole-exome sequencing (WES), facilitate the comprehensive analysis of an individual's genetic makeup, revealing rare variants, copy number variations, and structural rearrangements that may contribute to disease risk. Biomarker discovery efforts aim to identify genetic markers, gene expression signatures, or protein biomarkers that can predict disease progression, treatment response, or adverse drug reactions.

Practical Deployment of Genomic Profiling

Let's consider an example of genomic profiling using whole-exome sequencing data and the GATK (Genome Analysis Toolkit) software:

bashCopy code

```
gatk HaplotypeCaller -R reference_genome.fa -I input.bam -O output.vcf
```

In this command, "gatk HaplotypeCaller" performs variant calling on the aligned sequencing reads (input.bam) using a reference genome (reference_genome.fa). The output is stored in variant call format (VCF) file format (output.vcf), containing information about genetic variants identified in the individual's exome.

Clinical Applications of Personalized Medicine

Personalized medicine has transformative implications across various domains of clinical practice, including:

Cancer Treatment: Personalized oncology utilizes genomic profiling to identify actionable mutations, guide targeted therapies, and predict treatment response in cancer patients. Biomarker-driven therapies, such as tyrosine kinase inhibitors and immune checkpoint inhibitors, have revolutionized cancer treatment by improving patient outcomes and reducing treatment-related toxicity.

Pharmacogenomics: Pharmacogenomics aims to optimize drug therapy by considering patients' genetic variations that influence drug metabolism, efficacy, and toxicity. Pharmacogenetic testing enables clinicians to individualize drug dosing and selection, minimizing adverse drug reactions and maximizing therapeutic benefits.

Inherited Disorders: Personalized medicine offers new insights into the diagnosis and management of inherited disorders, such as rare genetic diseases and genetic predispositions to common diseases. Genetic counseling and predictive genetic testing empower individuals and families to make informed decisions about their health and medical care.

Preventive Medicine: Personalized risk assessment based on genetic and environmental factors enables early detection, risk stratification, and targeted interventions to prevent disease onset or progression. Lifestyle modifications, preventive screenings, and tailored interventions can mitigate disease risk and promote wellness in at-risk individuals.

Challenges and Future Directions

Despite its transformative potential, personalized medicine faces several challenges, including data privacy concerns, regulatory issues, and healthcare disparities. Future research directions may focus on addressing these challenges, advancing computational algorithms for data analysis and interpretation, and integrating multi-omics data to enhance predictive modeling and treatment optimization.

In summary, personalized medicine represents a paradigm shift in healthcare, harnessing the power of genomics and data-driven approaches to deliver more precise, effective, and patient-centered care. By integrating genomic profiling, biomarker discovery, and clinical decision support systems, personalized medicine holds promise for improving patient outcomes, reducing healthcare costs, and revolutionizing healthcare delivery in the 21st century.

BOOK 2
CODING IN BIOINFORMATICS
FROM SCRIPTING TO ADVANCED APPLICATIONS

ROB BOTWRIGHT

Chapter 1: Introduction to Programming for Bioinformatics

In the realm of bioinformatics, programming serves as the cornerstone for analyzing biological data, modeling biological systems, and developing computational tools and algorithms. This chapter embarks on an exploration of the diverse programming paradigms employed in bioinformatics, ranging from procedural and object-oriented programming to functional and declarative programming, unraveling their strengths, applications, and practical implementations in the field.

Understanding Programming Paradigms

Programming paradigms represent distinct approaches or styles of programming, each with its own set of principles, concepts, and methodologies. In bioinformatics, different programming paradigms are employed to address specific challenges and requirements in data analysis, algorithm design, and software development. These paradigms encompass procedural programming, object-oriented programming (OOP), functional programming, and declarative programming, each offering unique advantages and applications in bioinformatics.

Procedural Programming

Procedural programming is a fundamental programming paradigm that organizes code into procedures or functions, enabling step-by-step execution of algorithms and data manipulation tasks. In bioinformatics, procedural programming is often used for tasks such as data preprocessing, sequence alignment, and statistical analysis. Programming languages like Python, Perl, and C are commonly used for procedural programming in bioinformatics.

Practical Deployment of Procedural Programming

Let's consider an example of procedural programming in Python for calculating the GC content of DNA sequences:

pythonCopy code

```
def calculate_gc_content(sequence): gc_count = sequence.count('G') + sequence.count('C') total_bases = len(sequence) gc_content = (gc_count / total_bases) * 100 return gc_content dna_sequence = "ATCGATCGATCGATCG" gc_content = calculate_gc_content(dna_sequence) print("GC content:", gc_content)
```

In this Python script, the **calculate_gc_content** function computes the GC content of a DNA sequence by counting the occurrences of 'G' and 'C' nucleotides and dividing by the total number of bases.

Object-Oriented Programming (OOP)

Object-oriented programming (OOP) is a programming paradigm that models software components as objects with properties and behaviors, facilitating modular design, code reuse, and abstraction. In bioinformatics, OOP is utilized for developing software libraries, data structures, and graphical user interfaces (GUIs). Programming languages like Python, Java, and C++ support object-oriented programming paradigms in bioinformatics applications.

Practical Deployment of Object-Oriented Programming

Let's consider an example of object-oriented programming in Python for modeling DNA sequences as objects:

pythonCopy code

```
class DNASequence: def __init__(self, sequence):
self.sequence = sequence def
calculate_gc_content(self): gc_count =
self.sequence.count('G') + self.sequence.count('C')
total_bases = len(self.sequence) gc_content =
(gc_count / total_bases) * 100 return gc_content
dna_sequence =
DNASequence("ATCGATCGATCGATCG") gc_content =
dna_sequence.calculate_gc_content() print("GC
content:", gc_content)
```

In this Python script, the **DNASequence** class represents a DNA sequence object with methods for calculating the GC content.

Functional Programming

Functional programming is a programming paradigm that emphasizes the use of pure functions, immutability, and higher-order functions, enabling concise, modular, and expressive code. In bioinformatics, functional programming is employed for tasks such as data transformation, filtering, and map-reduce operations. Programming languages like Haskell, Scala, and Clojure support functional programming paradigms in bioinformatics applications.

Practical Deployment of Functional Programming

Let's consider an example of functional programming in Haskell for calculating the reverse complement of DNA sequences:

haskellCopy code

```haskell
reverseComplement :: String -> String
reverseComplement = reverse . map complement
where complement 'A' = 'T' complement 'T' = 'A'
complement 'C' = 'G' complement 'G' = 'C' complement
_ = error "Invalid nucleotide" main :: IO () main = do let
dnaSequence = "ATCGATCGATCGATCG" let
revComplement = reverseComplement dnaSequence
putStrLn $ "Reverse complement: " ++ revComplement
```

In this Haskell script, the **reverseComplement** function calculates the reverse complement of a DNA sequence using higher-order functions like **map** and **reverse**.

Declarative Programming

Declarative programming is a programming paradigm that emphasizes expressing the desired outcome rather than the step-by-step procedure for achieving it, enabling concise, readable, and maintainable code. In bioinformatics, declarative programming is utilized for tasks such as data querying, pattern matching, and rule-based inference. Domain-specific languages (DSLs) and query languages like SQL and XPath support declarative programming paradigms in bioinformatics applications.

Practical Deployment of Declarative Programming

Let's consider an example of declarative programming in SQL for querying DNA sequences from a database:

sqlCopy code

```
SELECT sequence FROM dna_sequences WHERE id =
'ABC123';
```

In this SQL query, the **SELECT** statement retrieves the DNA sequence corresponding to the specified identifier from the **dna_sequences** table.

Applications of Programming Paradigms in Bioinformatics

Programming paradigms play a crucial role in various bioinformatics applications, including:

Genomic Data Analysis: Procedural and functional programming are used for processing and analyzing large-scale genomic datasets, such as next-generation sequencing (NGS) data and microarray data.

Algorithm Development: Object-oriented programming facilitates the development of reusable software components and algorithms for sequence alignment, motif discovery, and phylogenetic analysis.

Software Engineering: Declarative programming enables the development of domain-specific languages (DSLs) and query languages for querying biological databases, designing biological models, and specifying analysis workflows.

Data Visualization: Object-oriented programming and functional programming paradigms are employed for developing interactive data visualization tools and graphical user interfaces (GUIs) for visualizing biological data and analysis results.

Challenges and Future Directions

While programming paradigms offer powerful tools for bioinformatics research and application development, challenges remain in terms of interoperability, scalability, and performance optimization across different paradigms. Future research directions may focus on developing hybrid approaches that combine the strengths of multiple programming paradigms, advancing programming languages and tools tailored to bioinformatics, and promoting interdisciplinary collaborations between computer scientists and biologists.

In summary, programming paradigms serve as versatile tools for tackling the complex challenges of bioinformatics, offering diverse approaches for data analysis, algorithm development, and software engineering. By understanding and harnessing the strengths of procedural, object-oriented, functional, and declarative programming paradigms, bioinformaticians can develop innovative solutions, advance scientific discoveries, and empower personalized healthcare in the era of genomic medicine.

The Significance of Programming Skills in Bioinformatics

In the interdisciplinary field of bioinformatics, where biology, computer science, and statistics converge, programming skills serve as the foundation upon which groundbreaking discoveries are made, innovative tools are developed, and complex biological questions are answered. This chapter delves into the critical importance of programming skills in bioinformatics, elucidating their pivotal role in data analysis, algorithm development, and scientific research, while also providing insights into how aspiring bioinformaticians can enhance their programming proficiency to thrive in this dynamic field.

The Intersection of Biology and Computer Science

Bioinformatics, at its core, bridges the gap between biology and computer science, leveraging computational tools and techniques to analyze biological data, model biological systems, and unravel the mysteries of life at the molecular level. Programming skills form the backbone of bioinformatics, enabling researchers to manipulate, analyze, and interpret vast amounts of biological data generated by high-throughput sequencing technologies, microarray experiments, and structural biology studies.

Data Analysis and Visualization

One of the primary applications of programming skills in bioinformatics is in data analysis and visualization. Through the use of programming languages like Python, R, and Perl, bioinformaticians can write scripts and algorithms to process raw biological data, perform statistical analyses, and visualize results in the form of graphs, charts, and interactive plots. From identifying differentially expressed genes in transcriptomic data to visualizing protein structures using molecular visualization tools, programming skills empower bioinformaticians to extract meaningful insights from complex biological datasets.

Practical Deployment of Programming Skills

Let's consider an example of using Python for data analysis and visualization in bioinformatics:

bashCopy code

python script.py input_data.txt output_plot.png

In this command, "python" executes the Python script named "script.py", which takes input data from the file "input_data.txt" and generates an output plot saved as "output_plot.png". Within the Python script, bioinformaticians can utilize libraries such as NumPy, pandas, and Matplotlib to manipulate data and create visualizations.

Algorithm Development and Optimization

Another crucial aspect of programming skills in bioinformatics is algorithm development and optimization. From sequence alignment algorithms to machine learning models for predictive modeling, bioinformaticians rely on programming to implement and refine algorithms that address biological questions and challenges. Proficiency in algorithm design and optimization allows bioinformaticians to develop efficient, scalable, and accurate computational methods for analyzing biological data and predicting biological phenomena.

Practical Deployment of Algorithm Development

Let's consider an example of implementing a basic sequence alignment algorithm in Python:

pythonCopy code

```
def sequence_alignment(sequence1, sequence2): #
Implementation of sequence alignment algorithm pass
alignment_result = sequence_alignment("ATCGATCG",
"ATAGATCG")          print("Alignment          result:",
alignment_result)
```

In this Python script, the **sequence_alignment** function implements a basic sequence alignment algorithm. Bioinformaticians can further optimize and extend this algorithm to handle more complex scenarios, such as global or local alignment, using dynamic programming techniques.

Tool Development and Automation

Programming skills are also essential for developing bioinformatics tools and pipelines that automate data analysis workflows, streamline repetitive tasks, and enable reproducible research. By leveraging programming languages and frameworks, bioinformaticians can create custom software tools, command-line utilities, and web applications tailored to specific research needs and requirements, enhancing productivity and collaboration within the scientific community.

Practical Deployment of Tool Development

Let's consider an example of developing a command-line tool in Python for processing FASTQ files:

```bash
python fastq_processor.py input.fastq output_processed.fastq
```

In this command, "fastq_processor.py" is a Python script that processes input FASTQ files and generates processed FASTQ files as output. Bioinformaticians can design and implement functionalities within the script to handle tasks such as quality control, read trimming, and sequence alignment, thereby creating a versatile and customizable tool for FASTQ file processing.

Education and Skill Development

Given the central role of programming skills in bioinformatics, education and skill development initiatives play a crucial role in nurturing the next generation of bioinformaticians. Bioinformatics training programs, workshops, and online resources offer opportunities for aspiring bioinformaticians to learn programming languages, algorithms, and software development practices relevant to the field. Hands-on projects, internships, and collaborative research experiences further reinforce programming skills and foster interdisciplinary collaboration within the bioinformatics community.

In summary, programming skills lie at the heart of bioinformatics, empowering researchers to analyze biological data, develop computational tools, and advance scientific knowledge at the intersection of biology and computer science. By honing their programming proficiency and embracing a mindset of continuous learning and innovation, bioinformaticians can unlock new insights into the complexities of life and contribute to transformative discoveries with profound implications for human health, agriculture, and the environment.

Chapter 2: Essential Tools and Languages for Bioinformatics Coding

In the dynamic field of bioinformatics, where biological data meets computational analysis, a myriad of tools and software are instrumental in unraveling the mysteries of the genome, proteome, and beyond. This chapter provides an in-depth overview of bioinformatics tools and software, exploring their diverse functionalities, applications, and practical deployment strategies, while also shedding light on how researchers can leverage these tools to address biological questions, analyze data, and advance scientific discovery.

Understanding the Role of Bioinformatics Tools

Bioinformatics tools serve as indispensable resources for researchers, empowering them to process, analyze, and interpret biological data generated by high-throughput sequencing, microarray experiments, structural biology studies, and other experimental techniques. These tools encompass a wide range of functionalities, including sequence alignment, genome assembly, motif discovery, protein structure prediction, pathway analysis, and more. By harnessing the power of bioinformatics tools, researchers can uncover hidden patterns, elucidate biological mechanisms, and accelerate the pace of discovery in fields such as genomics, transcriptomics, proteomics, and metabolomics.

Exploring Categories of Bioinformatics Tools

Bioinformatics tools can be categorized based on their functionalities and applications, encompassing:

Sequence Analysis Tools: Tools for analyzing DNA, RNA, and protein sequences, including sequence alignment tools (e.g., BLAST, Bowtie), motif discovery tools (e.g., MEME, HOMER), and sequence annotation tools (e.g., ANNOVAR, SnpEff).

Structural Bioinformatics Tools: Tools for predicting and analyzing the three-dimensional structures of proteins and nucleic acids, including protein structure prediction tools (e.g., SWISS-MODEL, Phyre2), molecular docking tools (e.g., AutoDock, Vina), and molecular visualization tools (e.g., PyMOL, UCSF Chimera).

Genomic Analysis Tools: Tools for analyzing genome-wide data, including genome assembly tools (e.g., SPAdes, Velvet), variant calling tools (e.g., GATK, Samtools), and genome annotation tools (e.g., MAKER, AUGUSTUS).

Transcriptomic Analysis Tools: Tools for analyzing gene expression data, including differential expression analysis tools (e.g., DESeq2, edgeR), pathway analysis tools (e.g., DAVID, GSEA), and alternative splicing analysis tools (e.g., MISO, rMATS).

Proteomic and Metabolomic Analysis Tools: Tools for analyzing protein and metabolite data, including mass spectrometry data analysis tools (e.g., ProteoWizard, XCMS), pathway enrichment analysis tools (e.g., MetaboAnalyst, MetScape), and protein-protein interaction prediction tools (e.g., STRING, APID).

Practical Deployment of Bioinformatics Tools

Let's consider an example of using the BLAST (Basic Local Alignment Search Tool) for sequence similarity searching:

bashCopy code

```
blastn -query input_sequence.fasta -db nr -out output_results.txt -evalue 0.001 -num_alignments 10
```

In this command, "blastn" performs a nucleotide sequence similarity search using the input sequence provided in the FASTA format ("-query") against the NCBI non-redundant (nr) database ("-db"). The results are saved in the "output_results.txt" file, with a specified E-value cutoff ("-evalue") and maximum number of alignments ("-num_alignments").

Choosing the Right Tool for the Task

Selecting the appropriate bioinformatics tool for a given task requires careful consideration of factors such as the type of data, the research question, computational resources, and user expertise. Researchers should evaluate the functionalities, accuracy, performance, and user-friendliness of different tools before making a decision. Online resources such as bioinformatics tool databases, user reviews, and tutorials can provide valuable insights into the strengths and limitations of various tools, helping researchers make informed choices that align with their research objectives and constraints.

Challenges and Future Directions

While bioinformatics tools have greatly facilitated biological research and discovery, challenges remain in terms of interoperability, reproducibility, usability, and scalability. Future directions in bioinformatics tool development may focus on enhancing integration with data repositories, standardizing data formats and workflows, improving user interfaces and documentation, and leveraging emerging technologies such as cloud computing, artificial intelligence, and high-performance computing to address these challenges and empower researchers with more powerful and accessible tools for biological data analysis.

In summary, bioinformatics tools and software play a pivotal role in advancing our understanding of the biological world, enabling researchers to analyze, interpret, and derive insights from vast amounts of biological data. By navigating the landscape of bioinformatics tools, researchers can leverage diverse functionalities and applications to address biological questions, explore new frontiers of discovery, and ultimately contribute to advancements in fields such as medicine, agriculture, biotechnology, and environmental science.

Selecting the Right Programming Language for Bioinformatics

In the realm of bioinformatics, where computational analysis meets biological data, the choice of programming language plays a crucial role in enabling researchers to manipulate, analyze, and interpret biological data effectively. This chapter provides a comprehensive comparison of programming languages commonly used in bioinformatics, highlighting their strengths, weaknesses, and suitability for different tasks, while also offering guidance on how researchers can navigate the landscape of programming languages to select the most appropriate tool for their specific needs.

Understanding the Landscape of Programming Languages

Bioinformatics encompasses a diverse range of tasks and applications, including sequence analysis, structural bioinformatics, genomic analysis, transcriptomic analysis, and more. Consequently, researchers often rely on a variety of programming languages to address different challenges and requirements in their research projects. Some of the most commonly used programming languages in bioinformatics include Python, R, Perl, Java, C/C++, and Julia, each offering unique features, libraries, and ecosystems tailored to the needs of bioinformaticians.

Python: Versatility and Ease of Use

Python has emerged as one of the most popular programming languages in bioinformatics due to its versatility, ease of use, and extensive library ecosystem. With libraries such as Biopython, NumPy, pandas, and Matplotlib, Python enables bioinformaticians to perform a wide range of tasks, including sequence analysis, data manipulation, statistical analysis, and visualization. Its simple syntax, readability, and cross-platform compatibility make it an ideal choice for beginners and experienced programmers alike, facilitating rapid prototyping, development, and deployment of bioinformatics tools and pipelines.

Practical Deployment of Python

Let's consider an example of using Python for sequence alignment with the Biopython library:

bashCopy code

```
python align_sequences.py input_sequences.fasta
```

In this command, "align_sequences.py" is a Python script that uses the Biopython library to perform sequence alignment on input sequences provided in the FASTA format. The script may utilize functions such as **SeqIO.read()** to read sequences from the input file and **pairwise2.align.globalxx()** to perform global sequence alignment with a scoring matrix.

R: Statistical Analysis and Data Visualization

R is another popular programming language in bioinformatics, particularly well-suited for statistical analysis, data visualization, and exploratory data analysis. With packages such as Bioconductor, DESeq2, and ggplot2, R empowers bioinformaticians to analyze high-dimensional biological data, perform differential expression analysis, visualize genomic features, and generate publication-quality plots and graphics. Its rich ecosystem of packages and active community support make it an invaluable tool for researchers working in fields such as genomics, transcriptomics, and proteomics.

Practical Deployment of R

Let's consider an example of using R for differential expression analysis:

bashCopy code

```
Rscript     analyze_expression.R     input_counts.csv
output_results.csv
```

In this command, "analyze_expression.R" is an R script that performs differential expression analysis on gene expression counts provided in the input file "input_counts.csv". The script may utilize packages such as DESeq2 to model gene expression data, perform statistical tests, and identify differentially expressed genes. The results are saved in the "output_results.csv" file for further analysis and interpretation.

Perl: Text Processing and Scripting

Perl, known for its robust text processing capabilities and powerful scripting features, has been a staple programming language in bioinformatics for many years. Its concise syntax, regular expression support, and extensive library of bioinformatics modules make it well-suited for tasks such as file parsing, data manipulation, and batch processing. While Perl's popularity has waned in recent years due to the rise of Python and R, it remains a valuable tool for bioinformaticians with existing Perl-based workflows and scripts.

Practical Deployment of Perl

Let's consider an example of using Perl for batch processing of sequence files:

bashCopy code

```
perl     process_sequences.pl     input_directory/
output_directory/
```

In this command, "process_sequences.pl" is a Perl script that processes sequence files located in the input directory "input_directory/" and generates output files in the specified output directory "output_directory/". The script may utilize Perl modules such as BioPerl to parse sequence files, extract sequence features, and perform sequence analysis tasks.

Choosing the Right Programming Language

Selecting the right programming language for a bioinformatics project depends on various factors, including the nature of the data, the complexity of the analysis, computational requirements, user expertise, and community support. Researchers should evaluate the strengths and weaknesses of different programming languages in terms of performance, scalability, ease of use, and availability of relevant libraries and tools. Experimentation, prototyping, and collaboration with peers can help researchers identify the most suitable programming language for their specific research needs and objectives.

In summary, programming languages play a pivotal role in bioinformatics, enabling researchers to analyze biological data, develop computational tools, and advance scientific knowledge at the interface of biology and computer science. By understanding the strengths and capabilities of different programming languages, bioinformaticians can navigate the landscape of programming languages effectively, selecting the most appropriate tool for their specific research projects and contributing to the advancement of bioinformatics research and discovery.

Chapter 3: Scripting Basics: Automating Tasks in Bioinformatics

Scripting languages serve as indispensable tools in the arsenal of bioinformaticians, offering a flexible and efficient means of automating tasks, processing data, and developing custom solutions tailored to the unique challenges of biological research. This chapter provides an introductory exploration of scripting languages in bioinformatics, delving into their key characteristics, advantages, and practical applications, while also equipping readers with essential knowledge and skills to harness the full potential of scripting languages in their research endeavors.

Understanding Scripting Languages

Scripting languages, such as Python, Perl, and Bash, are high-level programming languages designed to facilitate the rapid development of scripts or programs that automate tasks and manipulate data. Unlike compiled languages like C/C++ or Java, which require explicit compilation before execution, scripting languages are interpreted at runtime, allowing for immediate feedback and iterative development. This inherent flexibility makes scripting languages well-suited for prototyping, exploratory analysis, and ad-hoc scripting tasks in bioinformatics.

Exploring the Role of Scripting Languages in Bioinformatics

In the context of bioinformatics, scripting languages play a multifaceted role, enabling researchers to perform a wide range of tasks, including:

Data Parsing and Manipulation: Scripting languages excel at parsing and manipulating various types of biological data, such as sequence files, alignment results, and genomic annotations. By leveraging built-in libraries and modules, bioinformaticians can extract relevant information, filter data, and transform formats to suit their analysis needs.

Workflow Automation: Scripting languages streamline repetitive tasks and workflows in bioinformatics by automating data processing, analysis pipelines, and result generation. By writing scripts that orchestrate the execution of multiple tools and processes, researchers can save time, reduce errors, and increase reproducibility in their research projects.

Tool Development and Integration: Scripting languages enable the development of custom bioinformatics tools, utilities, and workflows tailored to specific research objectives and requirements. By integrating existing software tools and libraries or developing new functionalities from scratch, bioinformaticians can address specialized research challenges and advance scientific discovery.

Practical Deployment of Scripting Languages

Let's consider an example of using Python for parsing and analyzing sequence data:

bashCopy code

```
python sequence_analysis.py input_sequences.fasta
```

In this command, "sequence_analysis.py" is a Python script that reads input sequences from a FASTA file ("input_sequences.fasta") and performs sequence analysis tasks, such as calculating sequence lengths, identifying motifs, or generating sequence statistics. The script may utilize Python libraries such as Biopython for sequence manipulation and analysis.

Choosing the Right Scripting Language

Selecting the most appropriate scripting language for a bioinformatics task depends on factors such as familiarity, performance requirements, library availability, and community support. Python, with its simplicity, versatility, and extensive library ecosystem, has emerged as a popular choice among bioinformaticians for general-purpose scripting tasks. Perl, renowned for its powerful text processing capabilities and bioinformatics-specific modules, remains a viable option for tasks requiring advanced pattern matching and string manipulation. Bash, a shell scripting language, is well-suited for system administration tasks, file management, and command-line interface (CLI) automation in bioinformatics workflows.

Education and Skill Development

Mastering scripting languages in bioinformatics requires continuous learning, practice, and experimentation. Bioinformatics training programs, online tutorials, and community forums provide valuable resources and support for researchers looking to enhance their scripting skills. By engaging in hands-on projects, collaborating with peers, and exploring real-world datasets, bioinformaticians can deepen their understanding of scripting languages and unlock new possibilities for innovation and discovery in their research endeavors.

In summary, scripting languages are indispensable tools in the toolkit of bioinformaticians, offering a versatile and efficient means of automating tasks, processing data, and developing custom solutions in the pursuit of biological knowledge. By embracing scripting languages such as Python, Perl, and Bash, researchers can streamline workflows, accelerate analyses, and unlock new insights into the complexities of life at the molecular level. With dedication, creativity, and a solid foundation in scripting languages, bioinformaticians can navigate the evolving landscape of biological data analysis and contribute to transformative discoveries with profound implications for human health, agriculture, and the environment.

Empowering Bioinformatics with Scripting for Data Processing and Analysis

Scripting has become an indispensable tool in the field of bioinformatics, offering researchers the ability to automate data processing and analysis tasks efficiently. This chapter explores the role of scripting in bioinformatics, delving into its applications, benefits, and practical deployment strategies for handling and analyzing biological data effectively.

Automating Data Processing with Scripting

In bioinformatics, researchers often deal with large volumes of biological data generated from various sources, including next-generation sequencing, microarray experiments, and structural biology studies. Scripting provides a powerful mechanism for automating the processing of these datasets, allowing researchers to perform tasks such as data cleaning, formatting, and normalization with ease and efficiency.

Practical Deployment of Scripting for Data Processing

Let's consider an example of using Python to preprocess next-generation sequencing data:

bashCopy code

```
python    preprocess_data.py    input_reads.fastq
output_processed_reads.fastq
```

In this command, "preprocess_data.py" is a Python script designed to preprocess next-generation sequencing reads stored in the FASTQ format. The script reads the input reads from the file "input_reads.fastq," performs preprocessing steps such as quality trimming and adapter removal, and writes the processed reads to the file "output_processed_reads.fastq."

Facilitating Data Analysis with Scripting

Once data is processed, the next step in bioinformatics research involves analyzing the data to extract meaningful insights. Scripting plays a crucial role in this phase by enabling researchers to implement algorithms, perform statistical analyses, and visualize results. Whether it's identifying differentially expressed genes in transcriptomic data or predicting protein structures from amino acid sequences, scripting provides the flexibility and power needed to tackle diverse analytical tasks in bioinformatics.

Practical Deployment of Scripting for Data Analysis

Let's consider an example of using R to analyze gene expression data:

bashCopy code

```
Rscript analyze_gene_expression.R input_counts.csv output_results.csv
```

In this command, "analyze_gene_expression.R" is an R script designed to analyze gene expression data stored in the CSV format. The script reads the input counts from the file "input_counts.csv," performs differential expression analysis using statistical methods such as DESeq2, and writes the analysis results to the file "output_results.csv."

Benefits of Scripting in Bioinformatics

Scripting offers several advantages for data processing and analysis in bioinformatics:

Flexibility: Scripting languages such as Python, R, and Perl provide a high degree of flexibility, allowing researchers to customize scripts to their specific analysis needs.

Automation: By automating data processing and analysis tasks, scripting helps researchers save time and effort, enabling them to focus on higher-level research questions.

Reproducibility: Scripts can be easily shared and reused, promoting reproducible research practices and facilitating collaboration within the scientific community.

Scalability: Scripting languages are well-suited for handling large datasets and scaling analysis workflows to meet the demands of modern biological research.

Integration: Scripts can be integrated with existing bioinformatics tools and pipelines, enhancing their functionality and interoperability.

Challenges and Considerations

While scripting offers numerous benefits for data processing and analysis in bioinformatics, there are also challenges and considerations to keep in mind. These include:

Learning Curve: Mastering scripting languages and programming concepts may require time and effort, particularly for researchers with limited programming experience.

Performance: Depending on the complexity of the analysis, scripting solutions may not always offer the same performance as compiled languages or specialized bioinformatics tools.

Maintenance: Scripts may require regular updates and maintenance to ensure compatibility with changing data formats, software dependencies, and analysis requirements.

Documentation: Proper documentation of scripts is essential for reproducibility and transparency in research. Researchers should document their scripts thoroughly to facilitate understanding and reuse by others.

In summary, scripting plays a crucial role in bioinformatics by enabling researchers to automate data processing and analysis tasks effectively. By leveraging scripting languages such as Python, R, and Perl, researchers can streamline workflows, accelerate analyses, and extract valuable insights from biological data. With the right skills, tools, and practices in place, scripting empowers bioinformaticians to tackle complex research challenges and drive innovation in the field of bioinformatics.

Chapter 4: Data Manipulation and Parsing Techniques

In the realm of bioinformatics, where vast amounts of biological data are generated and analyzed, the efficient management and manipulation of data are paramount. This chapter explores the role of data structures in bioinformatics, delving into their significance, types, and practical deployment strategies for handling diverse biological data effectively.

Understanding the Importance of Data Structures

Data structures serve as the foundation for organizing and storing biological data in a structured and efficient manner. By choosing appropriate data structures, bioinformaticians can optimize data access, retrieval, and manipulation operations, ultimately enhancing the performance and scalability of bioinformatics algorithms and applications.

Types of Data Structures in Bioinformatics

Sequences: DNA, RNA, and protein sequences are fundamental entities in bioinformatics, often represented using string data structures. Additionally, specialized data structures such as suffix trees, suffix arrays, and Burrows-Wheeler Transform (BWT) matrices are employed for efficient sequence indexing and searching tasks.

Graphs: Graph data structures are used to represent biological networks, such as protein-protein interaction networks, metabolic pathways, and phylogenetic trees. Graph algorithms enable bioinformaticians to analyze network topology, identify functional modules, and infer evolutionary relationships among biological entities.

Matrices: Matrices are widely used to represent biological data, such as sequence alignments, expression matrices, and genomic variant tables. Sparse matrix representations optimize storage and computational efficiency for large-scale data analysis tasks.

Trees: Tree data structures, such as phylogenetic trees and hierarchical clustering dendrograms, are employed to represent hierarchical relationships among biological entities. Tree traversal algorithms enable bioinformaticians to explore evolutionary relationships, classify organisms, and infer ancestral sequences.

Practical Deployment of Data Structures

Let's consider an example of using a suffix array data structure for sequence searching:

bashCopy code

```
samtools faidx reference_genome.fasta
```

In this command, "samtools faidx" generates an index file for the reference genome in the FASTA format, creating a suffix array data structure that enables fast sequence retrieval and searching operations.

Benefits of Data Structures in Bioinformatics

Efficiency: Well-designed data structures optimize memory usage and computational performance, enabling bioinformaticians to analyze large-scale biological datasets efficiently.

Scalability: Scalable data structures accommodate the growing volume and complexity of biological data, ensuring that bioinformatics algorithms and applications remain viable and effective over time.

Flexibility: Data structures provide a flexible framework for representing diverse types of biological data, supporting a wide range of bioinformatics tasks and applications.

Interoperability: Standardized data structures facilitate interoperability and data exchange between different bioinformatics tools and platforms, enabling seamless integration and collaboration within the scientific community.

Challenges and Considerations

Data Representation: Choosing the appropriate data structure depends on the nature of the biological data and the specific analysis requirements. Bioinformaticians must carefully evaluate different data structures to ensure optimal performance and accuracy.

Algorithm Design: Effective algorithm design relies on selecting data structures that support efficient data access and manipulation operations. Bioinformaticians must consider algorithmic complexity and computational resources when designing bioinformatics algorithms and workflows.

Data Integrity: Data structures must accurately represent biological phenomena and preserve the integrity of the underlying biological data. Bioinformaticians must validate data structures to ensure their reliability and consistency in bioinformatics analyses.

In summary, data structures are essential components of bioinformatics, enabling bioinformaticians to organize, store, and manipulate biological data effectively. By leveraging appropriate data structures, bioinformaticians can optimize the performance, scalability, and accuracy of bioinformatics algorithms and applications, ultimately advancing our understanding of biological systems and driving innovation in the field of bioinformatics. With a deep understanding of data structures and their practical deployment strategies, bioinformaticians can tackle complex research challenges and unlock new insights into the mysteries of life at the molecular level.

Parsing Biological Data Formats: Techniques and Strategies

In the domain of bioinformatics, the ability to parse and manipulate diverse biological data formats is essential for conducting effective data analysis and interpretation. This chapter explores various techniques and strategies for parsing biological data formats, encompassing a wide range of file formats commonly encountered in bioinformatics research. From FASTA and FASTQ sequences to genomic annotations and alignment files, mastering the art of parsing these formats is crucial for bioinformaticians seeking to unlock the insights hidden within biological data.

Understanding the Landscape of Biological Data Formats

Biological data comes in a myriad of formats, each tailored to specific types of biological information and analysis tasks. Common biological data formats include:

FASTA: A simple text-based format for representing nucleotide or protein sequences, characterized by header lines starting with '>' followed by sequence data.

FASTQ: A text-based format for representing sequence data along with quality scores, typically used in next-generation sequencing experiments.

SAM/BAM: Formats for representing sequence alignment data, with SAM being the human-readable text format and BAM being the binary compressed version.

GFF/GTF: Formats for representing genomic annotations, including gene structures, transcripts, and other genomic features.

VCF: Variant Call Format, used for representing genomic variants such as single nucleotide polymorphisms (SNPs) and insertions/deletions (indels).

Parsing Techniques for Biological Data Formats

Text Processing: For simple formats like FASTA and FASTQ, text processing techniques such as regular expressions or string manipulation functions in programming languages like Python or Perl can be used to extract sequence data and metadata from files.

Parser Libraries: Many programming languages offer specialized libraries for parsing specific biological data formats. For instance, Biopython provides modules for parsing FASTA, FASTQ, and other sequence-related formats, while pysam offers tools for working with SAM/BAM files in Python.

File Format Converters: In some cases, converting data to a more standardized format can simplify parsing. Tools like **samtools** and **bedtools** offer command-line utilities for converting between different genomic file formats, making it easier to work with data in downstream analyses.

Structured Data Processing: For more complex formats like GFF/GTF and VCF, parsing often involves processing structured data using specialized libraries or tools. For instance, libraries like PyVCF in Python provide utilities for parsing and manipulating VCF files, while Bioconductor offers packages for working with genomic annotations in R.

Practical Deployment of Parsing Techniques

Let's consider an example of parsing a FASTA file using Biopython in Python:

bashCopy code

```
python parse_fasta.py input.fasta
```

In this command, "parse_fasta.py" is a Python script that utilizes the Biopython library to parse sequences from a FASTA file ("input.fasta"). The script reads the input file, extracts sequence data and metadata, and performs any desired analysis or processing tasks.

Best Practices for Parsing Biological Data Formats

Error Handling: Implement robust error handling mechanisms to gracefully handle unexpected data formats or corruption issues.

Efficiency: Optimize parsing algorithms for efficiency, particularly when dealing with large-scale datasets to minimize computational overhead.

Documentation: Document parsing scripts and workflows thoroughly to enhance reproducibility and facilitate collaboration with other researchers.

Validation: Validate parsed data against known standards or reference datasets to ensure accuracy and consistency in analysis results.

In summary, parsing biological data formats is a fundamental skill for bioinformaticians, enabling them to extract valuable insights from the wealth of biological data available. By employing a combination of text processing techniques, specialized libraries, and structured data processing tools, bioinformaticians can effectively parse a wide range of biological data formats, facilitating downstream analysis and interpretation. With careful attention to best practices and efficient parsing strategies, bioinformaticians can unlock the full potential of biological data and advance our understanding of complex biological systems.

Chapter 5: File Handling and Input/Output Operations

File handling constitutes a foundational aspect of bioinformatics programming, enabling the reading, writing, and manipulation of biological data stored in various formats. This chapter explores the nuances of file handling in bioinformatics programming, elucidating essential techniques, best practices, and practical deployment strategies for proficiently managing biological data files through command-line interface (CLI) commands and programming languages.

Understanding the Importance of File Handling

Effective file handling is indispensable in bioinformatics programming, where researchers routinely encounter diverse file formats such as FASTA, FASTQ, SAM/BAM, GFF/GTF, and VCF. Mastery of file handling techniques empowers bioinformaticians to access, extract, and analyze biological data efficiently, facilitating a wide array of computational tasks ranging from sequence alignment to variant analysis.

Essential Techniques for File Handling

Reading and Writing Files: Bioinformatics programming languages like Python, Perl, and R offer built-in functions and libraries for reading from and writing to files. For instance, in Python, the **open()** function is used to open files for reading or writing, while the **write()** method is employed to write data to a file.

Parsing File Formats: File parsing involves extracting relevant information from structured data files, such as sequences from FASTA files or alignment records from

SAM/BAM files. Specialized libraries and modules, such as Biopython for Python or Bioconductor for R, streamline the parsing process by providing functions tailored to specific file formats.

Error Handling: Robust error handling mechanisms are essential for gracefully handling exceptions and errors that may occur during file handling operations. Techniques such as try-except blocks in Python enable bioinformaticians to catch and handle errors effectively, ensuring the reliability and robustness of their programs.

Practical Deployment of File Handling Techniques

Let's consider an example of reading sequences from a FASTA file using Python:

bashCopy code

```
python read_fasta.py input.fasta
```

In this command, "read_fasta.py" is a Python script that utilizes the Biopython library to parse sequences from a FASTA file ("input.fasta"). The script reads the input file, extracts sequence data and metadata, and performs any desired analysis or processing tasks.

Best Practices for File Handling

Resource Management: Close files after reading from or writing to them to release system resources and avoid memory leaks. In Python, the **close()** method is used to close a file object.

Input Validation: Validate input files to ensure they conform to expected formats and standards, reducing the risk of errors and inaccuracies in subsequent analyses.

File Organization: Adopt a systematic approach to file naming and organization to enhance clarity and maintainability of bioinformatics projects. Consistent naming conventions and directory structures facilitate collaboration and reproducibility.

Documentation: Document file handling operations and workflows comprehensively to facilitate understanding and reuse by other researchers. Clear documentation improves the reproducibility and transparency of bioinformatics analyses.

In summary, file handling is a fundamental aspect of bioinformatics programming, enabling bioinformaticians to manage, manipulate, and analyze biological data effectively. By mastering essential file handling techniques and best practices, bioinformaticians can navigate the complexities of biological data files with confidence, streamlining computational workflows and accelerating scientific discoveries. With proficiency in file handling, bioinformatics programmers can unlock the full potential of biological data and contribute to advancements in fields such as genomics, proteomics, and structural biology.

Mastering Input and Output Operations for Biological Data

In the realm of bioinformatics, effective management of biological data is crucial for successful analysis and interpretation. This chapter delves into the intricacies of input and output (I/O) operations for biological data, exploring essential techniques, best practices, and practical deployment strategies for handling biological data through command-line interface (CLI) commands and programming languages.

Understanding the Significance of I/O Operations

Input and output operations are fundamental components of bioinformatics workflows, enabling researchers to access, process, and store biological data efficiently. Proficiency in I/O operations is essential for bioinformaticians seeking to navigate the diverse landscape of biological data formats and effectively leverage computational tools and algorithms for data analysis.

Essential Techniques for I/O Operations

Reading Biological Data: Reading biological data from files is a common task in bioinformatics programming. Command-line utilities such as **cat** and **grep** enable users to view and search text-based data files, while programming languages like Python, Perl, and R offer libraries and modules for reading structured biological data formats such as FASTA, FASTQ, SAM/BAM, and VCF.

Writing Biological Data: Writing biological data to files facilitates data storage and sharing, enabling researchers to document analysis results and archive experimental data. CLI commands like **echo** and **printf** allow users to create and write data to text files, while programming languages provide functions and methods for writing data to various file formats.

Parsing and Formatting: Parsing and formatting techniques are essential for extracting relevant information from biological data files and representing data in a structured and readable format. Specialized libraries and modules, such as Biopython for Python or Bioconductor for R, streamline the parsing process by providing functions tailored to specific biological data formats.

Practical Deployment of I/O Operations

Let's consider an example of reading sequences from a FASTA file using the **cat** command in the CLI:

bashCopy code

```
cat sequences.fasta
```

In this command, **cat** is used to display the contents of the file "sequences.fasta," which contains nucleotide or protein sequences in the FASTA format.

Best Practices for I/O Operations

Data Validation: Validate input data to ensure it conforms to expected formats and standards, reducing the risk of errors and inaccuracies in subsequent analyses.

Error Handling: Implement robust error handling mechanisms to gracefully handle exceptions and errors that may occur during I/O operations. Techniques such as try-except blocks in Python enable bioinformaticians to catch and handle errors effectively, ensuring the reliability and robustness of their programs.

Resource Management: Close files after reading from or writing to them to release system resources and avoid memory leaks. In programming languages like Python, the **close()** method is used to close a file object.

In summary, input and output operations are essential components of bioinformatics workflows, enabling bioinformaticians to access, process, and store biological data efficiently. By mastering essential I/O techniques and best practices, bioinformaticians can navigate the complexities of biological data formats with confidence, streamlining computational workflows and accelerating scientific discoveries. With proficiency in I/O operations, bioinformaticians can unlock the full potential of biological data and contribute to advancements in fields such as genomics, proteomics, and structural biology.

Chapter 6: Intermediate Coding Concepts in Bioinformatics

Control flow structures form the backbone of bioinformatics programming, providing the means to direct the flow of execution and make decisions based on biological data. This chapter explores the intricacies of control flow structures in bioinformatics programming, elucidating essential techniques, best practices, and practical deployment strategies through command-line interface (CLI) commands and programming languages.

Understanding the Significance of Control Flow Structures

Control flow structures enable bioinformaticians to orchestrate complex computational workflows, iterate over datasets, and conditionally execute code based on biological data properties. Mastery of control flow structures is essential for bioinformaticians seeking to manipulate and analyze biological data effectively, facilitating tasks such as sequence alignment, variant analysis, and gene expression profiling.

Essential Control Flow Structures

Conditional Statements: Conditional statements allow bioinformaticians to execute code based on specific conditions or criteria. Common conditional statements include **if, elif,** and **else**, which enable branching based on the evaluation of Boolean expressions. These statements are invaluable for implementing decision-making logic in bioinformatics algorithms and workflows.

Loops: Loops enable bioinformaticians to iterate over datasets, perform repetitive tasks, and process large volumes of biological data efficiently. Common types of loops include **for** loops, which iterate over a sequence of items, and **while** loops, which execute code as long as a specified condition remains true. Loops are essential for tasks such as sequence alignment, sequence assembly, and gene expression analysis.

Exception Handling: Exception handling mechanisms enable bioinformaticians to gracefully handle errors and exceptions that may occur during program execution. Techniques such as **try**, **except**, and **finally** enable bioinformaticians to catch and handle errors, ensuring the robustness and reliability of bioinformatics programs.

Practical Deployment of Control Flow Structures

Let's consider an example of using conditional statements to filter sequence data based on length in Python:

bashCopy code

```
python filter_sequences.py input.fasta output_filtered.fasta
```

In this command, "filter_sequences.py" is a Python script that reads sequences from the input FASTA file ("input.fasta"), filters sequences based on a specified length criterion using conditional statements, and writes the filtered sequences to the output FASTA file ("output_filtered.fasta").

Best Practices for Control Flow Structures

Clarity and Readability: Write control flow structures in a clear and readable manner to enhance code comprehension and maintainability. Use meaningful variable names, comments, and whitespace to improve code readability.

Efficiency: Optimize control flow structures for efficiency, particularly when processing large-scale biological datasets to minimize computational overhead and execution time.

Error Handling: Implement robust error handling mechanisms to gracefully handle errors and exceptions that may occur during program execution. Use exception handling constructs such as **try** and **except** to catch and handle errors effectively, ensuring the reliability and robustness of bioinformatics programs.

In summary, control flow structures are fundamental components of bioinformatics programming, enabling bioinformaticians to direct the flow of execution, make decisions based on biological data properties, and handle errors and exceptions effectively. By mastering essential control flow structures and adhering to best practices, bioinformaticians can develop robust and efficient bioinformatics algorithms and workflows, accelerating scientific discoveries and advancing our understanding of complex biological systems. With proficiency in control flow structures, bioinformaticians can unlock the full potential of biological data and contribute to advancements in fields such as genomics, proteomics, and structural biology.

Navigating Error Handling and Debugging Techniques in Bioinformatics Programming

Error handling and debugging are essential skills for bioinformaticians, enabling them to identify and resolve issues in their code efficiently. This chapter explores the nuances of error handling and debugging techniques in bioinformatics programming, providing insights into essential strategies, best practices, and practical deployment strategies through command-line interface (CLI) commands and programming languages.

Understanding the Importance of Error Handling and Debugging

Error handling and debugging play a crucial role in bioinformatics programming, where accuracy and reliability are paramount. By mastering error handling and debugging techniques, bioinformaticians can identify and resolve issues in their code, ensuring the robustness and integrity of bioinformatics analyses and applications.

Essential Error Handling Techniques

Exception Handling: Exception handling mechanisms enable bioinformaticians to gracefully handle errors and exceptions that may occur during program execution. In Python, for example, **try**, **except**, and **finally** blocks allow bioinformaticians to catch and handle exceptions, ensuring the robustness of their code.

Error Reporting: Effective error reporting mechanisms facilitate the identification and resolution of issues in bioinformatics programs. Techniques such as logging and error messages enable bioinformaticians to communicate errors and debugging information to users, enhancing the usability and maintainability of bioinformatics applications.

Practical Deployment of Error Handling Techniques

Let's consider an example of using exception handling to handle errors in a Python script:

bashCopy code

```
python analyze_data.py input.txt
```

In this command, "analyze_data.py" is a Python script that analyzes data from the input file ("input.txt"). If an error occurs during data analysis, exception handling mechanisms within the script catch and handle the error, ensuring the robustness of the program.

Debugging Techniques

Print Statements: Print statements are a simple yet effective debugging technique, allowing bioinformaticians to inspect variable values and track the flow of execution in their code. By strategically placing print statements throughout their code, bioinformaticians can identify issues and understand program behavior more effectively.

Interactive Debuggers: Interactive debuggers provide advanced debugging capabilities, allowing bioinformaticians to step through their code, inspect variable values, and analyze program behavior in real-time. Tools such as pdb for Python and RStudio for R offer interactive debugging features, enhancing the efficiency and effectiveness of debugging workflows.

Best Practices for Error Handling and Debugging

Documentation: Document error handling strategies and debugging techniques comprehensively to facilitate understanding and reuse by other bioinformaticians. Clear documentation improves the reproducibility and transparency of bioinformatics analyses, enabling users to understand and troubleshoot code effectively.

Testing: Implement comprehensive testing strategies to validate the functionality and reliability of bioinformatics programs. Techniques such as unit testing and integration testing enable bioinformaticians to identify and address issues in their code early in the development process, reducing the risk of errors in production environments.

In summary, error handling and debugging are essential skills for bioinformaticians, enabling them to identify and resolve issues in their code efficiently. By mastering essential error handling techniques and debugging strategies, bioinformaticians can ensure the robustness and reliability of bioinformatics analyses and applications, accelerating scientific discoveries and advancing our understanding of complex biological systems. With proficiency in error handling and debugging, bioinformaticians can unlock the full potential of bioinformatics programming and contribute to advancements in fields such as genomics, proteomics, and structural biology.

Chapter 7: Object-Oriented Programming for Bioinformatics

Object-Oriented Programming (OOP) is a powerful paradigm that revolutionizes software development by organizing code into reusable, modular units called objects. This chapter dives into the fundamental principles of OOP, elucidating essential concepts, practical applications, and deployment strategies relevant to bioinformatics programming.

Understanding the Significance of OOP

OOP offers a structured approach to software development, promoting code reusability, modularity, and maintainability. By encapsulating data and behavior within objects, OOP facilitates the creation of complex systems with clear hierarchies and relationships, making it particularly well-suited for bioinformatics applications where complex data structures and algorithms abound.

Essential OOP Concepts

Classes and Objects: Classes serve as blueprints for creating objects, defining their attributes (properties) and behaviors (methods). Objects are instances of classes, representing specific entities or concepts within a program. In Python, for instance, classes are defined using the **class** keyword, and objects are instantiated using constructor methods such as **__init__()**.

Encapsulation: Encapsulation refers to the bundling of data and methods within objects, shielding them from external interference and ensuring data integrity. By encapsulating data and behavior, OOP promotes code modularity and reusability, facilitating the development of robust and maintainable bioinformatics applications.

Inheritance: Inheritance allows classes to inherit attributes and methods from parent classes, enabling code reuse and hierarchy creation. Subclasses inherit properties and behaviors from superclasses, allowing for specialization and extension of functionality. In bioinformatics, inheritance is commonly used to model relationships between biological entities, such as genes, proteins, and sequences.

Polymorphism: Polymorphism enables objects of different classes to be treated interchangeably, facilitating flexibility and extensibility in bioinformatics programming. Polymorphic behavior allows methods to behave differently depending on the object they operate on, enabling the development of versatile and adaptable bioinformatics applications.

Practical Deployment of OOP Concepts

Let's consider an example of defining a **Sequence** class in Python for representing biological sequences:

bashCopy code

python sequence.py

In this command, "sequence.py" is a Python script that defines a **Sequence** class with attributes such as **id** and **sequence**, along with methods for reading sequences from files and performing basic sequence manipulation operations.

Best Practices for OOP

Modularity: Design classes with a clear and focused purpose, adhering to the single responsibility principle to ensure code modularity and maintainability. Well-designed classes are reusable and extensible, facilitating the development of complex bioinformatics systems.

Abstraction: Use abstraction to hide implementation details and expose only relevant interfaces to users. Abstraction enables bioinformaticians to focus on high-level concepts and functionality, reducing complexity and enhancing code readability.

Testing: Implement comprehensive testing strategies to validate the functionality and reliability of object-oriented code. Unit tests, integration tests, and mock objects enable bioinformaticians to identify and address issues early in the development process, ensuring the robustness and reliability of bioinformatics applications.

In summary, Object-Oriented Programming (OOP) is a powerful paradigm that revolutionizes bioinformatics programming by promoting code reusability, modularity, and maintainability. By mastering essential OOP concepts such as classes, objects, encapsulation, inheritance, and polymorphism, bioinformaticians can develop robust and scalable bioinformatics applications, accelerating scientific discoveries and advancing our understanding of complex biological systems. With proficiency in OOP, bioinformaticians can unlock the full potential of bioinformatics programming and contribute to advancements in fields such as genomics, proteomics, and structural biology.

Implementing Object-Oriented Programming (OOP) in Bioinformatics Applications

Object-Oriented Programming (OOP) is a powerful paradigm widely employed in bioinformatics for developing scalable, modular, and maintainable applications. This chapter delves into the practical aspects of implementing OOP in bioinformatics, elucidating essential techniques, best practices, and deployment strategies through command-line interface (CLI) commands and programming examples.

Understanding the Role of OOP in Bioinformatics

OOP provides a structured approach to software development, enabling bioinformaticians to model complex biological systems and processes effectively. By encapsulating data and behavior within objects, OOP facilitates code reuse, modularity, and extensibility, making it well-suited for bioinformatics applications where flexibility and scalability are paramount.

Essential Techniques for Implementing OOP in Bioinformatics

Class Design: Designing classes that accurately represent biological entities and processes is fundamental to successful OOP implementation in bioinformatics. Classes should encapsulate relevant data and behavior, adhering to principles such as the single responsibility principle and abstraction to ensure modularity and maintainability.

Inheritance and Polymorphism: Leveraging inheritance and polymorphism enables bioinformaticians to model hierarchical relationships and achieve code reuse. By defining base classes with common attributes and behaviors, bioinformaticians can create specialized subclasses that inherit and extend functionality as needed, promoting code extensibility and flexibility.

Encapsulation and Abstraction: Encapsulation and abstraction play a crucial role in OOP implementation, enabling bioinformaticians to hide implementation details and expose only relevant interfaces to users. By encapsulating data and behavior within objects and abstracting away unnecessary complexities, bioinformaticians can develop intuitive and user-friendly bioinformatics applications.

Practical Deployment of OOP in Bioinformatics

Let's consider an example of implementing OOP in a bioinformatics application using Python:

bashCopy code

python bioapp.py

In this command, "bioapp.py" is a Python script that implements a bioinformatics application using OOP principles. The script defines classes for representing biological sequences, performs sequence analysis tasks such as alignment and mutation detection, and generates informative outputs for further analysis.

Best Practices for OOP Implementation in Bioinformatics

Modularity and Reusability: Design classes with a clear and focused purpose, promoting code modularity and reusability. Well-designed classes are easier to maintain and extend, facilitating the development of complex bioinformatics applications.

Documentation and Testing: Document classes and methods comprehensively to facilitate understanding and reuse by other bioinformaticians. Implement comprehensive testing strategies to validate the functionality and reliability of OOP code, ensuring robustness and correctness.

Performance Optimization: Optimize OOP code for performance, particularly in computationally intensive bioinformatics applications. Techniques such as caching, parallelization, and algorithmic optimizations can enhance the efficiency and scalability of OOP-based bioinformatics applications.

In summary, implementing Object-Oriented Programming (OOP) in bioinformatics applications enables bioinformaticians to develop scalable, modular, and maintainable software solutions. By leveraging OOP principles such as class design, inheritance, encapsulation, and abstraction, bioinformaticians can model complex biological systems and processes effectively, accelerating scientific discoveries and advancing our understanding of life sciences. With proficiency in OOP implementation, bioinformaticians can unlock the full potential of bioinformatics programming and contribute to advancements in fields such as genomics, proteomics, and structural biology.

Chapter 8: Advanced Scripting Techniques and Best Practices

Optimizing bioinformatics scripts is essential for improving performance, reducing execution time, and enhancing the efficiency of computational workflows. This chapter explores various optimization techniques tailored to bioinformatics scripts, elucidating essential strategies, best practices, and practical deployment methods through command-line interface (CLI) commands and programming examples.

Understanding the Importance of Optimization

Optimization plays a crucial role in bioinformatics, where computational tasks often involve processing large datasets and executing complex algorithms. By optimizing bioinformatics scripts, bioinformaticians can maximize computational resources, minimize resource usage, and accelerate the execution of computational workflows, ultimately enhancing productivity and enabling more efficient data analysis.

Essential Optimization Techniques

Algorithmic Optimization: Algorithmic optimization involves refining algorithms and data structures to improve efficiency and reduce computational complexity. Techniques such as dynamic programming, memoization, and greedy algorithms can significantly enhance the performance of bioinformatics scripts, particularly in tasks such as sequence alignment, motif discovery, and graph algorithms.

Code Profiling: Code profiling allows bioinformaticians to identify performance bottlenecks and optimize critical sections of code effectively. Tools such as cProfile for Python and profilers for other programming languages enable bioinformaticians to analyze script execution times, memory usage, and function call frequencies, providing valuable insights into areas for optimization.

Parallelization: Parallelization involves distributing computational tasks across multiple processors or cores to execute them concurrently, thereby reducing execution time and enhancing scalability. Techniques such as multithreading and multiprocessing enable bioinformaticians to parallelize computationally intensive tasks, accelerating data processing and analysis in bioinformatics scripts.

Practical Deployment of Optimization Techniques

Let's consider an example of optimizing a bioinformatics script using algorithmic optimization techniques:

bashCopy code

```
python optimize_script.py
```

In this command, "optimize_script.py" is a Python script that implements a bioinformatics algorithm. By refining the algorithm using techniques such as dynamic programming or memoization, bioinformaticians can optimize the script to improve performance and reduce execution time.

Best Practices for Optimization

Prioritize Hotspots: Focus optimization efforts on critical sections of code (hotspots) identified through code profiling. Optimizing hotspots can yield significant performance improvements with minimal effort, maximizing the impact of optimization efforts.

Iterative Optimization: Adopt an iterative approach to optimization, making incremental improvements to bioinformatics scripts based on performance profiling and benchmarking results. Iterative optimization enables bioinformaticians to fine-tune scripts gradually, achieving optimal performance over time.

Documentation and Testing: Document optimization strategies and changes comprehensively to facilitate understanding and replication by other bioinformaticians. Implement comprehensive testing strategies to validate the functionality and reliability of optimized scripts, ensuring robustness and correctness.

In summary, optimization techniques are essential for enhancing the performance and efficiency of bioinformatics scripts, enabling bioinformaticians to maximize computational resources and accelerate data analysis. By leveraging algorithmic optimization, code profiling, parallelization, and other optimization strategies, bioinformaticians can optimize bioinformatics scripts effectively, improving productivity and enabling more efficient computational workflows. With proficiency in optimization techniques, bioinformaticians can unlock the full potential of bioinformatics programming and contribute to advancements in fields such as genomics, proteomics, and structural biology.

Best Practices for Efficient Scripting in Bioinformatics

Efficient scripting is indispensable in bioinformatics, where computational tasks often involve processing large datasets and executing complex algorithms. This chapter delves into the best practices for writing efficient scripts tailored to bioinformatics applications, elucidating essential strategies, deployment techniques, and CLI commands for optimizing scripting workflows.

Understanding the Importance of Efficiency

Efficiency is paramount in bioinformatics scripting, as it directly impacts productivity, resource utilization, and data analysis throughput. By adhering to best practices for efficient scripting, bioinformaticians can maximize computational resources, minimize execution time, and streamline data analysis workflows, ultimately accelerating scientific discoveries and advancing our understanding of biological systems.

Essential Best Practices for Efficient Scripting

Code Optimization: Optimizing code for performance is essential for efficient scripting in bioinformatics. Techniques such as algorithmic optimization, code profiling, and parallelization can significantly enhance the efficiency of bioinformatics scripts, reducing execution time and resource usage.

Modularity and Reusability: Designing scripts with modularity and reusability in mind promotes code maintainability and extensibility. By breaking down scripts into modular components and encapsulating functionality within functions or classes, bioinformaticians can reuse code across multiple projects, saving time and effort.

Documentation and Comments: Comprehensive documentation and meaningful comments enhance script readability and facilitate understanding by other bioinformaticians. Documenting script functionality, input/output formats, and dependencies improves code transparency and usability, enabling smoother collaboration and knowledge sharing.

Error Handling and Debugging: Implementing robust error handling mechanisms and debugging techniques is crucial for detecting and resolving issues in bioinformatics scripts. Techniques such as exception handling, logging, and interactive debugging tools enhance script reliability and maintainability, minimizing downtime and troubleshooting efforts.

Practical Deployment of Best Practices

Let's consider an example of deploying best practices for efficient scripting in bioinformatics using a Python script:

bashCopy code

```
python efficient_script.py input.fasta output.txt
```

In this command, "efficient_script.py" is a Python script that performs a bioinformatics analysis on the input FASTA file ("input.fasta") and generates results in the output text file ("output.txt"). The script adheres to best practices for efficiency, including code optimization, modularity, documentation, and error handling.

CLI Commands for Optimization

Code Profiling: Use the **cProfile** module in Python to profile script execution and identify performance bottlenecks:

bashCopy code

```
python    -m    cProfile    -o    profile_results.prof
efficient_script.py input.fasta output.txt
```
Memory Profiling: Employ tools like **memory_profiler** to profile memory usage during script execution:
bashCopy code
```
python    -m    memory_profiler    efficient_script.py
input.fasta output.txt
```

Best Practices for Deployment

Version Control: Utilize version control systems such as Git to manage script revisions and facilitate collaboration among team members.

Continuous Integration: Implement continuous integration (CI) pipelines to automate script testing, validation, and deployment processes, ensuring code quality and reliability.

In summary, adhering to best practices for efficient scripting is essential for maximizing productivity and accelerating data analysis in bioinformatics. By optimizing code, promoting modularity and reusability, documenting scripts comprehensively, and implementing robust error handling mechanisms, bioinformaticians can develop efficient scripts that enhance productivity and facilitate scientific discoveries. With proficiency in efficient scripting practices and deployment techniques, bioinformaticians can unlock the full potential of bioinformatics programming and contribute to advancements in fields such as genomics, proteomics, and structural biology.

Chapter 9: High-Performance Computing and Parallel Processing

High-Performance Computing (HPC) is a cornerstone of modern scientific research, including bioinformatics. This chapter provides an introductory overview of HPC, exploring its significance, principles, deployment techniques, and CLI commands commonly used in bioinformatics applications.

Significance of High-Performance Computing

HPC refers to the use of advanced computing systems to solve complex computational problems efficiently and expediently. In bioinformatics, HPC plays a crucial role in analyzing vast amounts of biological data, performing computationally intensive tasks such as sequence alignment, molecular dynamics simulations, and genomic analysis. By harnessing the computational power of HPC systems, bioinformaticians can accelerate scientific discoveries, unravel complex biological processes, and address pressing challenges in fields such as drug discovery, personalized medicine, and evolutionary biology.

Principles of High-Performance Computing

At the core of HPC lie several key principles that govern its operation and effectiveness:

Parallelism: HPC systems leverage parallel processing techniques to execute multiple computational tasks concurrently, thereby maximizing resource utilization and performance. Parallelism can be achieved at various levels, including task-level parallelism, data-level parallelism, and instruction-level parallelism, depending on the nature of the computational problem and the architecture of the HPC system.

Scalability: Scalability refers to the ability of HPC systems to efficiently handle increasing workloads and data volumes as computational demands grow. Scalability is essential for accommodating the ever-expanding datasets and computational challenges encountered in bioinformatics, enabling seamless expansion of computational resources to meet evolving research needs.

Performance Optimization: Performance optimization involves tuning HPC systems and applications to achieve maximum computational efficiency and throughput. Techniques such as code optimization, algorithmic optimization, and hardware acceleration (e.g., GPU computing) are commonly employed to enhance performance and reduce execution times in bioinformatics HPC workflows.

Deployment Techniques for High-Performance Computing

Deploying HPC systems and applications in bioinformatics requires careful consideration of various factors, including hardware infrastructure, software tools, and computational workflows. Common deployment techniques include:

Cluster Computing: Cluster computing involves connecting multiple computers (nodes) together to form a single computational entity. Tools such as OpenMPI and Slurm are commonly used to manage cluster resources, schedule jobs, and distribute computational tasks across nodes in bioinformatics HPC environments.

Cloud Computing: Cloud computing provides on-demand access to scalable computational resources over the internet, offering flexibility, cost-effectiveness, and ease of deployment for bioinformatics HPC applications. Platforms such as Amazon Web Services (AWS), Google Cloud Platform (GCP), and Microsoft Azure offer HPC-specific services and resources tailored to bioinformatics research.

Distributed Computing: Distributed computing involves distributing computational tasks across multiple nodes or systems connected via a network. Techniques such as MapReduce and Apache Spark facilitate distributed data processing and analysis in bioinformatics HPC workflows, enabling efficient handling of large-scale datasets and parallel execution of bioinformatics algorithms.

CLI Commands for High-Performance Computing

Submitting Jobs to a Cluster: Use the **sbatch** command to submit batch jobs to a Slurm-managed cluster:

bashCopy code

```
sbatch script.sh
```

Monitoring Cluster Resources: Monitor cluster resource usage and job status using commands such as **squeue**, **sinfo**, and **sacct**:

bashCopy code

```
squeue sinfo sacct -u username
```

In summary, High-Performance Computing (HPC) is a transformative technology that revolutionizes scientific research, including bioinformatics. By harnessing the computational power of HPC systems and deploying efficient computational workflows, bioinformaticians can accelerate data analysis, drive innovation, and make significant contributions to fields such as genomics, proteomics, and systems biology. With proficiency in HPC principles, deployment techniques, and CLI commands, bioinformaticians can unlock the full potential of HPC in bioinformatics research and propel scientific discovery forward.

Parallel Processing Techniques for Bioinformatics Applications

Parallel processing techniques have become indispensable in bioinformatics for handling large-scale data analysis and computationally intensive tasks. This chapter explores the significance of parallel processing, principles underlying parallel computing, deployment techniques, and CLI commands commonly used in bioinformatics applications.

Significance of Parallel Processing

Parallel processing enables bioinformaticians to leverage multiple computing resources concurrently, accelerating data analysis and computation-intensive tasks. In bioinformatics, where datasets are vast and algorithms are complex, parallel processing plays a crucial role in enhancing computational efficiency, scalability, and throughput. By distributing computational tasks across multiple processors or cores, parallel processing enables bioinformaticians to tackle challenging research questions, unravel complex biological phenomena, and accelerate scientific discoveries.

Principles of Parallel Computing

At the heart of parallel computing lie several key principles that govern its operation and effectiveness:

Concurrency: Concurrency refers to the simultaneous execution of multiple computational tasks or processes. In parallel computing, concurrency enables bioinformaticians to exploit available computing resources efficiently, maximizing throughput and minimizing execution time.

Parallelism: Parallelism involves breaking down computational tasks into smaller subtasks that can be executed concurrently on multiple processing units. Parallelism can be achieved at various levels, including task-level parallelism, data-level parallelism, and instruction-level parallelism, depending on the nature of the computational problem and the architecture of the parallel computing system.

Scalability: Scalability refers to the ability of parallel computing systems to efficiently handle increasing workloads and data volumes as computational demands grow. Scalability is essential for accommodating the ever-expanding datasets and computational challenges encountered in bioinformatics, enabling seamless expansion of computational resources to meet evolving research needs.

Deployment Techniques for Parallel Processing

Deploying parallel processing techniques in bioinformatics requires careful consideration of various factors, including hardware infrastructure, software tools, and computational workflows. Common deployment techniques include:

Multithreading: Multithreading involves dividing computational tasks into smaller threads that can be executed concurrently within a single process. Tools such as Python's threading module and Java's Executor framework facilitate multithreaded programming in bioinformatics applications, enabling efficient utilization of multicore processors and enhancing computational throughput.

Multiprocessing: Multiprocessing involves distributing computational tasks across multiple processes, each running independently and utilizing its own memory space. Tools such as Python's multiprocessing module and the Message Passing Interface (MPI) enable bioinformaticians to parallelize computationally intensive tasks across multiple processors or nodes in a distributed computing environment, enhancing scalability and performance.

CLI Commands for Parallel Processing

Executing Python Script with Multiprocessing: Use the **multiprocessing** module in Python to parallelize script execution across multiple processes:

bashCopy code

```
python script.py
```

Running MPI Job: Use the **mpiexec** command to run MPI jobs in a distributed computing environment:

bashCopy code

```
mpiexec -n 4 script
```

In summary, parallel processing techniques are indispensable tools for bioinformaticians seeking to accelerate data analysis, enhance computational efficiency, and tackle complex research questions in bioinformatics. By leveraging concurrency, parallelism, and scalability principles, bioinformaticians can deploy parallel processing techniques effectively, enabling seamless parallel execution of computational tasks and accelerating scientific discoveries in fields such as genomics, proteomics, and systems biology. With proficiency in deployment techniques and CLI commands for parallel processing, bioinformaticians can unlock the full potential of parallel computing in bioinformatics research and propel scientific innovation forward.

Chapter 10: Developing Bioinformatics Applications and Tools

The Software Development Life Cycle (SDLC) is a systematic process used to develop high-quality software efficiently. In the context of bioinformatics applications, SDLC plays a critical role in ensuring the reliability, functionality, and usability of software tools used for data analysis, algorithm development, and scientific research. This chapter explores the SDLC phases, principles, deployment techniques, and CLI commands commonly used in bioinformatics software development.

Understanding the SDLC

The SDLC consists of several distinct phases, each contributing to the overall development process:

Requirements Gathering: In this initial phase, bioinformaticians collaborate with stakeholders to identify and document software requirements, including functional specifications, user requirements, and performance criteria. Tools such as JIRA and Trello facilitate requirements gathering and management, enabling effective communication and collaboration among team members.

System Design: In the system design phase, bioinformaticians create detailed architectural and design specifications based on the gathered requirements. Design documents may include system architecture diagrams, data flow diagrams, and interface designs, providing a blueprint for software implementation. Tools such as Lucidchart and Draw.io support system design activities, enabling bioinformaticians to visualize and communicate design decisions effectively.

Implementation: The implementation phase involves translating design specifications into executable code. Bioinformaticians write code, develop algorithms, and implement software features according to established design guidelines and coding standards. Programming languages such as Python, R, and Java are commonly used in bioinformatics software development, offering flexibility, performance, and a rich ecosystem of libraries and frameworks.

Testing: Testing is a crucial phase in the SDLC, where bioinformaticians verify and validate software functionality, correctness, and performance. Techniques such as unit testing, integration testing, and regression testing are employed to detect defects and ensure software quality. Tools such as pytest for Python and JUnit for Java support automated testing, enabling bioinformaticians to execute test cases efficiently and systematically.

Deployment: The deployment phase involves deploying software to production environments and making it available to end users. Bioinformaticians use deployment techniques such as containerization (e.g., Docker), virtualization (e.g., VirtualBox), and cloud computing platforms (e.g., AWS, GCP) to deploy bioinformatics applications effectively. CLI commands such as **docker run** and **gcloud app deploy** facilitate the deployment process, enabling bioinformaticians to deploy software quickly and reliably.

Maintenance: The maintenance phase involves ongoing support, bug fixes, and enhancements to deployed software. Bioinformaticians monitor software performance, address user feedback, and implement updates to ensure continued functionality and usability. Techniques such as version control (e.g., Git) and issue tracking systems (e.g., GitHub Issues) facilitate software maintenance activities, enabling bioinformaticians to manage software evolution effectively.

Principles of Effective SDLC in Bioinformatics

Effective SDLC in bioinformatics relies on several key principles:

Iterative Development: Adopt an iterative development approach, where software is developed incrementally, tested rigorously, and refined based on user feedback and changing requirements. Iterative development enables bioinformaticians to deliver functional software iteratively, reducing development cycle times and enhancing software quality.

Collaboration and Communication: Foster collaboration and communication among multidisciplinary teams, including bioinformaticians, biologists, software engineers, and domain experts. Effective communication ensures shared understanding of requirements, design decisions, and software implementation, facilitating smoother development and deployment processes.

Documentation: Document software requirements, design decisions, and implementation details comprehensively to ensure traceability, reproducibility, and maintainability. Clear and well-documented software artifacts enable bioinformaticians to understand, modify, and extend software effectively, reducing cognitive overhead and minimizing errors during software development and maintenance.

In summary, the Software Development Life Cycle (SDLC) is a systematic approach to developing high-quality software efficiently. In bioinformatics, effective SDLC practices are essential for developing reliable, functional, and user-friendly software tools used for data analysis, algorithm development, and scientific research. By following SDLC phases, principles, and best practices, bioinformaticians can streamline software development processes, improve software quality, and accelerate scientific discoveries in fields such as genomics, proteomics, and systems biology. With proficiency in SDLC techniques and CLI commands for software development and deployment, bioinformaticians can unlock the full potential of bioinformatics software development and contribute to advancements in biomedical research and healthcare.

Testing, Documentation, and Deployment of Bioinformatics Tools

Testing, documentation, and deployment are critical aspects of the software development process in bioinformatics. This chapter explores the importance of testing, documentation, and deployment techniques in ensuring the reliability, usability, and scalability of bioinformatics tools. We'll delve into testing methodologies, documentation practices, and deployment techniques, along with CLI commands commonly used in the process.

Testing Bioinformatics Tools

Testing is essential for identifying and rectifying defects, ensuring software functionality, and validating scientific results. In bioinformatics, where data accuracy and reproducibility are paramount, rigorous testing is indispensable. Common testing methodologies include:

Unit Testing: Unit testing involves testing individual components or units of code in isolation to ensure they function as intended. Tools such as pytest for Python and JUnit for Java facilitate unit testing in bioinformatics applications:

bashCopy code

```
pytest test_module.py
```

Integration Testing: Integration testing focuses on testing the interaction between different components or modules to ensure they integrate seamlessly. Integration tests simulate real-world scenarios and verify the interoperability of software components.

Regression Testing: Regression testing involves re-running previously executed tests to ensure that recent code changes haven't introduced new defects or regressions. Automated regression testing helps maintain software stability and reliability across successive releases.

Performance Testing: Performance testing assesses the scalability, responsiveness, and resource utilization of bioinformatics tools under different workloads.

Techniques such as load testing and stress testing evaluate software performance under varying conditions.

Documentation Practices

Comprehensive documentation is essential for facilitating software understanding, adoption, and maintenance. In bioinformatics, where data analysis workflows can be complex and domain-specific, clear documentation is indispensable. Documentation practices include:

API Documentation: Documenting application programming interfaces (APIs) enables users to understand the functionality, inputs, and outputs of bioinformatics tools. Tools such as Sphinx and Doxygen generate API documentation from source code comments:

```bash
bashCopy code
sphinx-build -b html docs/ build/
```

User Manuals: User manuals provide step-by-step instructions for installing, configuring, and using bioinformatics tools. Manuals should include usage examples, command-line options, and troubleshooting tips to assist users in effectively utilizing the software.

Tutorials and Examples: Tutorials and example datasets help users familiarize themselves with bioinformatics tools and demonstrate common use cases and workflows. Interactive tutorials and Jupyter notebooks facilitate hands-on learning and experimentation:

```bash
bashCopy code
jupyter notebook tutorial.ipynb
```

Deployment Techniques

Deploying bioinformatics tools involves making software accessible to end-users in a reliable and reproducible manner. Deployment techniques include:

Containerization: Containerization technologies such as Docker enable encapsulating bioinformatics tools and their dependencies into portable, self-contained containers. Docker CLI commands facilitate container creation and deployment:

bashCopy code

```
docker build -t my_tool . docker run my_tool
```

Virtualization: Virtualization platforms such as VirtualBox and VMware enable running bioinformatics tools in isolated virtual environments, ensuring compatibility and reproducibility across different host systems.

Cloud Deployment: Cloud computing platforms such as AWS, GCP, and Azure offer scalable infrastructure and services for deploying bioinformatics tools in the cloud. CLI commands provided by cloud providers facilitate resource provisioning, deployment, and management:

bashCopy code

```
aws ec2 create-instance ...
```

In summary, testing, documentation, and deployment are integral components of the software development process in bioinformatics. By adopting rigorous testing methodologies, comprehensive documentation practices, and effective deployment techniques, bioinformaticians can ensure the reliability, usability, and scalability of bioinformatics tools. With proficiency in CLI commands and best practices for testing, documentation, and deployment, bioinformaticians can develop and deploy high-quality software tools that drive scientific discoveries and advancements in fields such as genomics, proteomics, and systems biology.

BOOK 3
EXPLORING DATA SCIENCE IN BIOINFORMATICS
TECHNIQUES AND TOOLS FOR ANALYSIS

ROB BOTWRIGHT

Chapter 1: Introduction to Data Science in Bioinformatics

Data science has emerged as a cornerstone in bioinformatics, revolutionizing the way biological data is analyzed, interpreted, and utilized. This chapter provides an in-depth exploration of the role of data science in bioinformatics, covering key concepts, methodologies, tools, and applications. From data acquisition and preprocessing to advanced analytics and machine learning, data science techniques play a pivotal role in extracting meaningful insights from vast biological datasets.

Understanding Data Science in Bioinformatics

Data science encompasses a multidisciplinary approach to extracting knowledge and insights from data, employing techniques from statistics, computer science, and domain-specific fields such as biology and genetics. In bioinformatics, data science techniques are applied to a wide range of biological data types, including genomic sequences, gene expression profiles, protein structures, and clinical data. The primary goals of data science in bioinformatics include:

Data Acquisition: Data science techniques facilitate the acquisition of diverse biological data from various sources, including public databases, high-throughput experiments, and clinical studies. Command-line tools such as **wget** and **curl** are commonly used to download datasets from online repositories:

bashCopy code

```
wget https://example.com/dataset.zip
```

Data Preprocessing: Raw biological data often contains noise, errors, and missing values, necessitating preprocessing steps such as quality control, normalization, and feature extraction. Tools such as **Pandas** in Python and **dplyr** in R provide powerful data manipulation capabilities:

bashCopy code

```
pip install pandas
```

Exploratory Data Analysis (EDA): EDA involves visualizing and summarizing biological data to uncover patterns, trends, and relationships. Techniques such as data visualization, descriptive statistics, and clustering are employed to gain insights into the underlying structure of biological datasets.

Statistical Analysis: Statistical methods are used to analyze biological data, assess significance, and infer biological hypotheses. Techniques such as hypothesis testing, regression analysis, and survival analysis are applied to identify biomarkers, genetic variants, and disease associations.

Machine Learning: Machine learning algorithms are used to build predictive models, classify biological samples, and identify complex patterns in biological data. Techniques such as supervised learning, unsupervised learning, and deep learning are employed to address diverse bioinformatics tasks.

Applications of Data Science in Bioinformatics

Data science techniques find applications across various domains of bioinformatics, including:

Genomic Analysis: Data science enables the analysis of genomic sequences, identification of genetic variants, and prediction of gene functions. Techniques such as sequence alignment, variant calling, and genome annotation are applied to study genetic variation and evolutionary relationships.

Transcriptomics: Transcriptomic data analysis involves quantifying gene expression levels, identifying differentially expressed genes, and inferring regulatory networks. Data science techniques such as RNA-seq analysis, gene expression clustering, and pathway enrichment analysis are employed in transcriptomics studies.

Proteomics: Data science techniques are used to analyze protein sequences, predict protein structures, and characterize protein-protein interactions. Techniques such as protein folding prediction, protein docking, and functional annotation aid in understanding protein functions and interactions.

Metagenomics: Metagenomic data analysis involves studying microbial communities, identifying microbial species, and characterizing their functional profiles. Data science techniques such as metagenomic assembly, taxonomic classification, and functional annotation are applied in metagenomics studies.

Challenges and Future Directions

Despite its transformative potential, data science in bioinformatics faces several challenges, including data heterogeneity, scalability, and interpretability. Future directions in data science in bioinformatics include the integration of multi-omics data, development of interpretable machine learning models, and advancement of data sharing and reproducibility practices.

In summary, data science plays a pivotal role in bioinformatics, enabling the analysis, interpretation, and utilization of vast biological datasets. By leveraging data science techniques such as data acquisition, preprocessing, exploratory data analysis, statistical analysis, and machine learning, bioinformaticians can unlock insights into biological processes, disease mechanisms, and therapeutic targets. With the continued advancement of data science methodologies and tools, the future of bioinformatics holds promise for accelerating scientific discoveries and improving human health.

Role and Importance of Data Science in Biomedical Research

In the realm of biomedical research, data science has emerged as a transformative force, reshaping the landscape of scientific inquiry, discovery, and innovation. This chapter delves into the pivotal role and profound importance of data science in biomedical research, exploring its applications, methodologies, and contributions to advancing our understanding of human health and disease.

Understanding Data Science in Biomedical Research

Data science in biomedical research encompasses a broad spectrum of methodologies, tools, and techniques aimed at extracting insights from complex biomedical datasets. Leveraging computational algorithms, statistical models, and machine learning approaches, data scientists analyze diverse datasets ranging from genomic sequences and clinical records to imaging data and drug interactions. The primary objectives of data science in biomedical research include:

Data Integration: Data science facilitates the integration of heterogeneous biomedical data from disparate sources, enabling researchers to combine genomic, clinical, and experimental data to gain a comprehensive understanding of biological systems.

Data Exploration: Data scientists employ exploratory data analysis techniques to uncover patterns, trends, and correlations within biomedical datasets, providing valuable insights into disease mechanisms, biomarker discovery, and therapeutic targets.

Predictive Modeling: By building predictive models using machine learning algorithms, data scientists can forecast disease outcomes, identify risk factors, and personalize treatment strategies, paving the way for precision medicine approaches tailored to individual patients.

Knowledge Discovery: Data science plays a crucial role in knowledge discovery by mining large-scale biomedical databases, identifying novel associations, and elucidating biological pathways underlying disease processes.

Applications of Data Science in Biomedical Research

Data science finds diverse applications across various domains of biomedical research, including:

Genomics and Personalized Medicine: Data science techniques enable genomic analysis, variant interpretation, and pharmacogenomics studies, facilitating the development of personalized medicine approaches based on individuals' genetic profiles.

Clinical Informatics: Data science plays a key role in clinical informatics by analyzing electronic health records (EHRs), predicting patient outcomes, and optimizing healthcare delivery through data-driven decision-making.

Medical Imaging: Data science techniques are applied in medical imaging analysis for image segmentation, feature extraction, and disease diagnosis, enhancing the accuracy and efficiency of diagnostic imaging modalities such as MRI, CT, and PET.

Drug Discovery and Development: Data science accelerates drug discovery and development processes by analyzing chemical structures, predicting drug-target interactions, and screening compound libraries for therapeutic efficacy.

Challenges and Opportunities

Despite its transformative potential, data science in biomedical research faces several challenges, including data heterogeneity, privacy concerns, and ethical considerations. Addressing these challenges requires interdisciplinary collaboration, robust data governance frameworks, and adherence to ethical guidelines for data use and sharing.

Nevertheless, data science presents immense opportunities for driving biomedical research forward, from unraveling the complexities of human biology to advancing precision medicine and improving patient outcomes. By harnessing the power of data science, researchers can unlock new avenues for understanding disease mechanisms, discovering novel therapeutics, and ultimately transforming healthcare delivery.

In summary, data science occupies a central role in biomedical research, empowering scientists to analyze vast amounts of data, uncover hidden patterns, and translate findings into actionable insights. By integrating computational methods, statistical analysis, and machine learning techniques, data scientists contribute to the advancement of knowledge in biology, medicine, and healthcare. With continued innovation and collaboration, data science holds the promise of revolutionizing biomedical research, leading to breakthroughs that enhance human health and well-being.

Chapter 2: Exploratory Data Analysis Techniques

Exploratory Data Analysis (EDA) is a foundational practice in data science, providing insights into datasets through visualization and statistical analysis. This chapter explores the fundamental principles and techniques of EDA, illustrating its importance in understanding data distributions, identifying patterns, and informing subsequent analyses.

Understanding Exploratory Data Analysis

Exploratory Data Analysis is the process of examining and visualizing data to understand its underlying structure, characteristics, and relationships. It involves summarizing key features of the dataset, detecting outliers, and assessing data quality. EDA is typically performed at the initial stages of data analysis to gain insights and inform subsequent modeling or hypothesis testing.

Key Techniques of Exploratory Data Analysis

Summary Statistics: Summary statistics provide a concise overview of the dataset's central tendency, dispersion, and shape. Common summary statistics include mean, median, standard deviation, and quartiles. The **describe** function in Python's Pandas library computes summary statistics for numerical variables:

bashCopy code

python -m pip install pandas

bashCopy code

```bash
python -c "import pandas as pd; data = pd.read_csv('data.csv'); print(data.describe())"
```

Data Visualization: Data visualization techniques, such as histograms, box plots, and scatter plots, offer intuitive ways to visualize the distribution, variability, and relationships within the dataset. Libraries such as Matplotlib and Seaborn in Python provide powerful tools for creating visualizations:

bashCopy code

```bash
python -m pip install matplotlib seaborn
```

bashCopy code

```bash
python -c "import seaborn as sns; import pandas as pd; data = pd.read_csv('data.csv'); sns.pairplot(data)"
```

Correlation Analysis: Correlation analysis assesses the strength and direction of linear relationships between variables. The correlation coefficient, ranging from -1 to 1, indicates the degree of association between variables. The **corr** function in Pandas computes pairwise correlations between numerical columns:

bashCopy code

```bash
python -c "import pandas as pd; data = pd.read_csv('data.csv'); print(data.corr())"
```

Outlier Detection: Outliers are data points that deviate significantly from the rest of the dataset and can skew statistical analyses. Techniques such as box plots, scatter plots, and z-score analysis can help identify outliers visually or statistically:

bashCopy code

```
python -c "import seaborn as sns; import pandas as pd;
data          =          pd.read_csv('data.csv');
sns.boxplot(x=data['column'])"
```

Applications of Exploratory Data Analysis

Exploratory Data Analysis finds applications across various domains, including:

Finance: EDA helps financial analysts understand market trends, identify anomalies in financial data, and inform investment decisions.

Healthcare: EDA aids healthcare professionals in analyzing patient data, identifying risk factors for diseases, and optimizing treatment protocols.

Marketing: EDA assists marketers in segmenting customers, analyzing campaign performance, and identifying target demographics.

Environmental Science: EDA enables environmental scientists to analyze environmental datasets, detect pollution trends, and assess the impact of climate change.

Challenges and Best Practices

Despite its benefits, EDA poses several challenges, including data preprocessing, missing values, and selection bias. Best practices for effective EDA include:

Data Cleaning: Addressing missing values, handling outliers, and standardizing data formats are essential steps in preparing data for EDA.

Visualization: Choosing appropriate visualization techniques and creating clear, interpretable visualizations are critical for effective communication of insights.

Interpretation: Contextualizing findings within the domain of interest and avoiding overinterpretation of exploratory analyses are important for drawing meaningful conclusions.

In summary, Exploratory Data Analysis serves as a cornerstone in the data science workflow, providing a systematic approach to understanding and visualizing datasets. By leveraging summary statistics, data visualization, correlation analysis, and outlier detection techniques, data scientists can gain valuable insights into data distributions, patterns, and relationships. With its broad applicability across diverse domains, EDA plays a crucial role in informing decision-making, hypothesis generation, and subsequent analyses. Through adherence to best practices and careful interpretation of results, EDA empowers data scientists to extract actionable insights and drive data-driven decision-making in various fields.

Visualization and Summary Statistics for Biological Data

In the realm of bioinformatics, the visualization and summary statistics of biological data play a crucial role in uncovering patterns, trends, and insights that are essential for understanding biological processes and informing scientific discoveries. This chapter delves into the fundamental principles and techniques of visualizing and summarizing biological data, exploring their significance and applications in various areas of bioinformatics research.

Understanding Biological Data Visualization

Biological data encompasses a wide range of data types, including genomic sequences, gene expression profiles, protein structures, and clinical data. Effective visualization techniques enable researchers to explore these complex datasets, identify meaningful patterns, and communicate findings to a wider audience. Visualization techniques commonly employed in bioinformatics include:

Histograms and Density Plots: Histograms and density plots provide visual representations of the distribution of continuous variables, such as gene expression levels or sequence read counts. These plots enable researchers to assess data distributions, detect outliers, and identify potential data transformation needs.

bashCopy code

```
Rscript -e "data <- read.table('expression_data.txt', header=TRUE); hist(data$expression)"
```

Box Plots and Violin Plots: Box plots and violin plots are used to visualize the distribution of continuous variables across different groups or categories. These plots provide insights into the central tendency, variability, and skewness of the data within each group.

bashCopy code

```
Rscript -e "data <- read.table('expression_data.txt', header=TRUE); boxplot(data$expression ~ data$group)"
```

Scatter Plots: Scatter plots are effective for visualizing relationships between two continuous variables, such as gene expression levels or protein concentrations. Scatter plots can reveal correlations, trends, and clusters within the data.

```bash
Copy code
```
```
Rscript -e "data <- read.table('expression_data.txt',
header=TRUE); plot(data$gene1, data$gene2,
xlab='Gene 1', ylab='Gene 2')"
```

Heatmaps: Heatmaps provide visual representations of large datasets, such as gene expression matrices or sequence alignments. Heatmaps use color gradients to represent data values, making it easier to identify patterns, clusters, and outliers.

```bash
Copy code
```
```
Rscript -e "data <- read.table('expression_matrix.txt',
header=TRUE); heatmap(data)"
```

Summary Statistics for Biological Data

Summary statistics offer a quantitative overview of biological data, summarizing key features such as central tendency, dispersion, and variability. Common summary statistics used in bioinformatics include:

Mean: The mean represents the average value of a dataset and is calculated by summing all values and dividing by the number of observations.

```bash
Copy code
```
```
python -c "import numpy as np; data =
np.loadtxt('expression_data.txt'); print(np.mean(data))"
```

Median: The median represents the middle value of a dataset when arranged in ascending order. It is less affected by outliers compared to the mean.

```bash
Copy code
```
```
python -c "import numpy as np; data =
np.loadtxt('expression_data.txt');
print(np.median(data))"
```

Standard Deviation: The standard deviation measures the spread or dispersion of data points around the mean. It provides insights into the variability within the dataset.

bashCopy code

```
python -c "import numpy as np; data = np.loadtxt('expression_data.txt'); print(np.std(data))"
```

Correlation Coefficient: The correlation coefficient quantifies the strength and direction of the linear relationship between two variables. It ranges from -1 to 1, where 1 indicates a perfect positive correlation, -1 indicates a perfect negative correlation, and 0 indicates no correlation.

bashCopy code

```
Rscript -e "data <- read.table('expression_data.txt', header=TRUE); print(cor(data$gene1, data$gene2))"
```

Applications and Significance

Visualization and summary statistics are indispensable tools in bioinformatics research, aiding researchers in:

Gene Expression Analysis: Visualizing gene expression profiles and summarizing expression levels across different conditions or treatments.

Genomic Variation Analysis: Visualizing genomic variants and summarizing their frequencies, distributions, and associations with diseases.

Protein Structure Analysis: Visualizing protein structures and summarizing their properties, interactions, and functional annotations.

Clinical Data Analysis: Visualizing clinical data and summarizing patient demographics, disease outcomes, and treatment responses.

By effectively visualizing and summarizing biological data, researchers can gain valuable insights into complex biological systems, identify biomarkers, unravel disease mechanisms, and ultimately advance our understanding of human health and disease.

In summary, visualization and summary statistics are indispensable tools in bioinformatics research, enabling researchers to explore, analyze, and interpret complex biological datasets. By employing techniques such as histograms, box plots, scatter plots, and summary statistics, researchers can uncover patterns, trends, and associations within biological data, leading to insights that drive scientific discovery and innovation. With the continued advancement of computational tools and visualization techniques, the role of visualization and summary statistics in bioinformatics research will continue to expand, empowering researchers to unravel the complexities of biological systems and improve human health.

Chapter 3: Data Visualization in Bioinformatics

Data visualization is a powerful tool for conveying information, patterns, and insights hidden within datasets. Next, we explore the fundamental principles of data visualization, the techniques used to effectively communicate data, and the importance of visual clarity and interpretation.

Understanding Data Visualization

Data visualization is the graphical representation of data to facilitate understanding. It transforms raw data into visual elements such as charts, graphs, and maps, making complex information more accessible and understandable. Data visualization serves multiple purposes, including exploration, analysis, communication, and decision-making.

Key Principles of Data Visualization

Clarity and Simplicity: A fundamental principle of data visualization is clarity and simplicity. Visualizations should convey information clearly and concisely, avoiding unnecessary clutter or complexity. Simple designs with clear labels and minimal distractions enhance the readability and comprehension of visualizations.

Accuracy and Integrity: Data visualizations must accurately represent the underlying data without distorting or misrepresenting information. Proper labeling, axis scaling, and data encoding ensure the integrity of visualizations, preventing misinterpretation or bias.

Relevance and Context: Effective data visualizations provide relevant context to aid interpretation and decision-making. Adding annotations, captions, or explanatory notes contextualizes the data and helps users understand its significance within a broader context.

Consistency and Standards: Consistency in design elements, color schemes, and labeling conventions maintains coherence across visualizations and facilitates comparison. Adhering to established standards and best practices ensures consistency and enhances usability.

Techniques for Data Visualization

Bar Charts: Bar charts are effective for comparing discrete categories or groups by representing data as bars of varying lengths. They are commonly used to visualize categorical data or compare quantities across different groups.

bashCopy code

```
python -m pip install matplotlib
```

bashCopy code

```
python -c "import matplotlib.pyplot as plt; plt.bar(categories, values); plt.xlabel('Categories'); plt.ylabel('Values'); plt.title('Bar Chart')"
```

Line Charts: Line charts display data trends over time or across continuous variables by connecting data points with lines. They are useful for visualizing sequential data or identifying patterns and trends.

bashCopy code

```
python -c "import matplotlib.pyplot as plt;
plt.plot(x_values, y_values); plt.xlabel('Time');
plt.ylabel('Values'); plt.title('Line Chart')"
```

Scatter Plots: Scatter plots represent individual data points as dots on a two-dimensional plane, allowing for the visualization of relationships between two variables. They are valuable for identifying correlations, clusters, and outliers within the data.

bashCopy code

```
python -c "import matplotlib.pyplot as plt;
plt.scatter(x_values, y_values); plt.xlabel('Variable 1');
plt.ylabel('Variable 2'); plt.title('Scatter Plot')"
```

Heatmaps: Heatmaps use color gradients to represent data values in a matrix format, making it easier to visualize patterns, clusters, and trends within large datasets. They are commonly used in genomics, finance, and geographic analysis.

bashCopy code

```
python -c "import seaborn as sns;
sns.heatmap(data_matrix); plt.xlabel('Variables');
plt.ylabel('Observations'); plt.title('Heatmap')"
```

Applications and Importance

Data visualization plays a vital role in various fields, including:

Business and Marketing: Visualizing sales data, customer demographics, and market trends to inform business strategies and decision-making.

Science and Research: Analyzing experimental data, visualizing scientific phenomena, and communicating research findings to a broader audience.

Healthcare and Medicine: Visualizing patient data, medical imaging, and epidemiological trends to improve healthcare delivery and patient outcomes.

Education and Communication: Enhancing learning experiences and facilitating knowledge dissemination through interactive and engaging visualizations.

In summary, data visualization is an essential tool for exploring, analyzing, and communicating data-driven insights. By adhering to principles of clarity, accuracy, relevance, and consistency, data visualizations can effectively convey complex information and facilitate informed decision-making. Through the use of techniques such as bar charts, line charts, scatter plots, and heatmaps, data visualizations enable users to uncover patterns, trends, and relationships within datasets, ultimately driving innovation and progress across diverse domains. As the volume and complexity of data continue to grow, the importance of data visualization in extracting actionable insights and empowering data-driven decision-making will only continue to increase.

Tools and Techniques for Visualizing Biological Data

Visualizing biological data is essential for understanding complex biological systems, identifying patterns, and extracting meaningful insights. Next, we explore various tools and techniques used in bioinformatics to visualize diverse biological datasets, ranging from genomic sequences to protein structures.

Understanding Biological Data Visualization

Biological data encompasses a wide range of datasets, including genomic sequences, gene expression profiles, protein structures, and phylogenetic trees. Effective visualization techniques transform these complex datasets into visual representations, enabling researchers to explore, analyze, and interpret biological phenomena.

Key Visualization Tools and Techniques

Genomic Sequence Visualization: Genomic sequences are visualized using tools such as Genome Browser, IGV (Integrative Genomics Viewer), and UCSC Genome Browser. These tools allow researchers to visualize genomic features, such as genes, regulatory elements, and genetic variants, in the context of the entire genome.

bashCopy code

```
igv -g reference_genome.fa -b alignment.bam
```

Gene Expression Visualization: Gene expression data is visualized using tools like Heatmaps, Volcano Plots, and Principal Component Analysis (PCA) plots. Heatmaps display gene expression levels across samples, while Volcano Plots highlight differentially expressed genes. PCA plots visualize sample clustering based on gene expression profiles.

bashCopy code

```
Rscript -e "library(ggplot2); data <- read.table('expression_matrix.txt', header=TRUE); p <- ggplot(data, aes(x=PC1, y=PC2, color=sample_type)); p + geom_point()"
```

Protein Structure Visualization: Protein structures are visualized using software such as PyMOL, Chimera, and

VMD (Visual Molecular Dynamics). These tools allow researchers to visualize protein structures in 3D, analyze molecular interactions, and identify functional domains.

bashCopy code

```
pymol protein_structure.pdb
```

Phylogenetic Tree Visualization: Phylogenetic trees are visualized using tools like FigTree, iTOL (Interactive Tree Of Life), and Archaeopteryx. These tools enable researchers to visualize evolutionary relationships among species, analyze branch lengths, and annotate tree nodes.

bashCopy code

```
figtree tree.nexus
```

Advanced Visualization Techniques

Interactive Visualization: Tools like D3.js (Data-Driven Documents) and Plotly enable interactive visualization of biological data on web platforms. Interactive features such as tooltips, zooming, and filtering enhance user engagement and exploration of complex datasets.

bashCopy code

```
npm install d3
```

Network Visualization: Network visualization tools like Cytoscape and Gephi are used to visualize biological networks, such as protein-protein interactions and gene regulatory networks. These tools enable researchers to analyze network topology, identify network modules, and visualize node attributes.

bashCopy code

```
cytoscape network_data.txt
```

Spatial Visualization: Spatial visualization techniques, such as spatial transcriptomics and spatial proteomics, enable the visualization of biomolecules within cellular or tissue contexts. Tools like STARmap and SpatialDE facilitate the analysis and visualization of spatially resolved omics data.

bashCopy code

```
python -m pip install spatialde
```

Applications and Significance

Biological data visualization has numerous applications in various domains, including:

Genomics and Genetics: Visualizing genomic features, genetic variants, and gene expression patterns to understand genetic mechanisms underlying diseases and traits.

Structural Biology: Visualizing protein structures, molecular interactions, and ligand binding sites to elucidate protein function and design novel therapeutics.

Systems Biology: Visualizing biological networks, signaling pathways, and metabolic pathways to analyze complex biological systems and predict cellular behaviors.

Evolutionary Biology: Visualizing phylogenetic trees, sequence alignments, and evolutionary relationships to study species divergence, adaptation, and speciation.

In summary, tools and techniques for visualizing biological data are essential for understanding complex biological processes, elucidating molecular mechanisms, and informing scientific discoveries. By leveraging a diverse range of visualization tools and techniques, researchers can explore, analyze, and interpret biological datasets with precision and clarity. From genomic sequences to protein structures to phylogenetic trees, effective visualization enables researchers to uncover patterns, trends, and relationships within biological data, ultimately advancing our understanding of life's complexities and contributing to scientific innovation and discovery. As the field of bioinformatics continues to evolve, the development and application of advanced visualization tools and techniques will play a critical role in driving biological research forward.

Chapter 4: Statistical Methods for Data Analysis

Statistical inference plays a crucial role in bioinformatics by providing methods for making predictions, testing hypotheses, and drawing conclusions from biological data. Next, we delve into the principles of statistical inference in bioinformatics, exploring its importance, techniques, and applications in analyzing biological datasets.

Understanding Statistical Inference

Statistical inference involves drawing conclusions about a population based on sample data, taking into account uncertainty and variability inherent in the data. In bioinformatics, statistical inference is used to analyze biological data, such as gene expression levels, genetic variants, and protein interactions, to gain insights into biological processes and phenomena.

Key Concepts in Statistical Inference

Population and Sample: In statistical inference, a population refers to the entire set of individuals or objects under study, while a sample is a subset of the population from which data is collected. Statistical inference aims to make inferences about the population based on the characteristics of the sample.

Parameter Estimation: Parameter estimation involves estimating unknown parameters of a population based on sample data. Common estimation techniques include maximum likelihood estimation (MLE) and method of moments. In bioinformatics, parameters such as gene expression levels, mutation rates, and protein concentrations are often estimated from biological data.

Hypothesis Testing: Hypothesis testing is used to assess the significance of observed differences or relationships in data. The process involves formulating null and alternative hypotheses, selecting an appropriate test statistic, and calculating the p-value to determine the likelihood of observing the data under the null hypothesis. Common statistical tests used in bioinformatics include t-tests, chi-square tests, and ANOVA.

Confidence Intervals: Confidence intervals provide a range of values within which the true parameter value is likely to lie with a certain level of confidence. They are constructed based on sample data and provide a measure of uncertainty around parameter estimates. Confidence intervals are used to quantify the precision of parameter estimates and assess the reliability of statistical inference.

Statistical Techniques in Bioinformatics

Differential Expression Analysis: In gene expression analysis, statistical techniques such as t-tests, ANOVA, and linear models are used to identify genes that are differentially expressed between experimental conditions. These techniques help elucidate molecular mechanisms underlying biological processes and diseases.

bashCopy code

```
Rscript -e "library(DESeq2); counts <- read.table('expression_matrix.txt', header=TRUE); condition <- factor(c('control', 'treatment')); dds <-
```

```
DESeqDataSetFromMatrix(countData=counts,
colData=DataFrame(condition)); dds <- DESeq(dds); res
<- results(dds);"
```

Genome-Wide Association Studies (GWAS): GWAS involve testing the association between genetic variants and traits of interest. Statistical techniques such as logistic regression and linear mixed models are used to identify genetic variants associated with complex traits and diseases.

bashCopy code

```
plink --bfile genotype_data --pheno phenotype_data.txt
--assoc --out association_results
```

Survival Analysis: Survival analysis is used to analyze time-to-event data, such as patient survival times or disease recurrence. Statistical techniques such as Kaplan-Meier estimation and Cox proportional hazards regression are used to assess the impact of covariates on survival outcomes.

bashCopy code

```
Rscript    -e    "library(survival);    survival_data    <-
read.table('survival_data.txt', header
```

Hypothesis Testing and Confidence Intervals for Biological Data

Hypothesis testing and confidence intervals are fundamental statistical techniques used in bioinformatics to analyze and draw conclusions from biological data. Next, we explore the principles behind hypothesis testing and confidence intervals, their applications in bioinformatics, and how they are deployed using command-line tools and techniques.

Understanding Hypothesis Testing

Hypothesis testing is a statistical method used to make inferences about population parameters based on sample data. It involves formulating a null hypothesis (H0) and an alternative hypothesis (Ha), selecting an appropriate test statistic, calculating its probability under the null hypothesis, and interpreting the results to determine the significance of the observed data.

In bioinformatics, hypothesis testing is used to assess the significance of observed differences or associations in biological data. For example, researchers may use hypothesis testing to determine whether there is a significant difference in gene expression levels between experimental conditions or to assess the association between genetic variants and disease traits.

Deploying Hypothesis Testing

One commonly used command-line tool for hypothesis testing in bioinformatics is R, a programming language and environment for statistical computing and graphics. Researchers can use R to perform a wide range of hypothesis tests, including t-tests, chi-square tests, ANOVA, and non-parametric tests.

For example, to perform a two-sample t-test to compare the mean gene expression levels between two experimental groups, researchers can use the **t.test** function in R:

bashCopy code

```
Rscript            -e            "data            <-
read.table('gene_expression_data.txt',   header=TRUE);
```

```
result      <-      t.test(data$group1,      data$group2);
print(result);"
```
This command reads the gene expression data from a text file, performs a two-sample t-test between the two experimental groups (**group1** and **group2**), and prints the test result, including the test statistic, p-value, and confidence interval.

Understanding Confidence Intervals

A confidence interval is a range of values calculated from sample data that is likely to contain the true population parameter with a certain level of confidence (e.g., 95% confidence). Confidence intervals provide a measure of the precision and uncertainty associated with parameter estimates obtained from sample data.

In bioinformatics, confidence intervals are used to estimate the range of plausible values for population parameters such as mean gene expression levels, effect sizes, and odds ratios. Researchers can use confidence intervals to assess the reliability and precision of their estimates and to make informed decisions based on the uncertainty associated with their data.

Deploying Confidence Intervals

To calculate confidence intervals for biological data, researchers can use statistical software packages such as R, Python with libraries like NumPy and SciPy, or command-line tools specifically designed for confidence interval estimation.

For example, to calculate a 95% confidence interval for the mean gene expression level in a dataset using Python and SciPy, researchers can use the **t.interval** function:

bashCopy code

```
python -c "import numpy as np; from scipy.stats import
t; data = np.loadtxt('gene_expression_data.txt'); mean =
np.mean(data); std_error = np.std(data) /
np.sqrt(len(data)); confidence_interval = t.interval(0.95,
len(data) - 1, loc=mean, scale=std_error);
print(confidence_interval);"
```

This command reads the gene expression data from a text file, calculates the mean and standard error of the data, and then uses the t-distribution to calculate the 95% confidence interval for the mean gene expression level.

Applications in Bioinformatics

Hypothesis testing and confidence intervals are widely used in bioinformatics for various applications, including:

Gene Expression Analysis: Assessing the significance of differential gene expression between experimental conditions.

Variant Discovery: Evaluating the association between genetic variants and disease traits in genome-wide association studies.

Protein Interaction Analysis: Testing the significance of protein-protein interactions and identifying functional modules within biological networks.

Phylogenetic Inference: Estimating confidence intervals for branch lengths and divergence times in phylogenetic trees.

In summary, hypothesis testing and confidence intervals are essential statistical techniques used in bioinformatics to analyze and draw conclusions from biological data. By deploying these techniques using command-line tools and statistical software packages such as R and Python, researchers can assess the significance of observed differences or associations in biological datasets, estimate population parameters with uncertainty, and make informed decisions based on the reliability of their estimates. Hypothesis testing and confidence intervals play a critical role in advancing our understanding of complex biological systems and driving biomedical research forward

Chapter 5: Machine Learning Fundamentals in Bioinformatics

Machine learning has emerged as a powerful tool in bioinformatics, offering techniques for extracting valuable insights from large and complex biological datasets. Next, we explore the principles of machine learning, its applications in bioinformatics, and how researchers can deploy machine learning algorithms using command-line tools and techniques.

Understanding Machine Learning

Machine learning is a subfield of artificial intelligence that focuses on developing algorithms and models that enable computers to learn from data and make predictions or decisions without being explicitly programmed. Machine learning algorithms can be broadly categorized into supervised learning, unsupervised learning, and reinforcement learning.

In bioinformatics, machine learning is used to analyze biological data, such as gene expression profiles, DNA sequences, protein structures, and biomedical images, to uncover patterns, identify relationships, and make predictions about biological phenomena.

Deploying Machine Learning Algorithms

One commonly used command-line tool for machine learning in bioinformatics is scikit-learn, a Python library that provides a wide range of machine learning algorithms and tools for data preprocessing, model training, evaluation, and prediction.

For example, to train a support vector machine (SVM) classifier to predict the functional class of genes based on their expression profiles, researchers can use the following command-line script:

bashCopy code

```
python -m pip install scikit-learn python -c "from sklearn import svm; from sklearn.model_selection import train_test_split; from sklearn.metrics import accuracy_score; import numpy as np; data = np.loadtxt('gene_expression_data.txt'); X = data[:, :-1]; y = data[:, -1]; X_train, X_test, y_train, y_test = train_test_split(X, y, test_size=0.2, random_state=42); clf = svm.SVC(); clf.fit(X_train, y_train); y_pred = clf.predict(X_test); accuracy = accuracy_score(y_test, y_pred); print('Accuracy:', accuracy);"
```

This command installs scikit-learn using pip, reads gene expression data from a text file, splits the data into training and testing sets, trains an SVM classifier on the training data, and evaluates its performance on the testing data using accuracy as the evaluation metric.

Applications of Machine Learning in Bioinformatics

Gene Expression Analysis: Machine learning algorithms are used to classify genes based on their expression patterns, identify biomarkers associated with diseases, and predict patient outcomes from gene expression profiles.

Sequence Analysis: Machine learning techniques such as hidden Markov models, neural networks, and random forests are used to predict gene regulatory elements, protein-protein interactions, and protein functions from DNA and protein sequences.

Structural Biology: Machine learning algorithms are applied to predict protein structures, analyze protein-ligand interactions, and design new drugs and therapeutics.

Clinical Decision Support: Machine learning models are used to analyze electronic health records, predict patient diagnoses and treatment outcomes, and assist clinicians in making personalized treatment decisions.

Challenges and Future Directions

Despite its potential, machine learning in bioinformatics faces several challenges, including the need for large and high-quality datasets, interpretability of models, and ethical considerations related to data privacy and bias. However, ongoing advances in machine learning algorithms, data integration techniques, and computational resources hold promise for addressing these challenges and unlocking new opportunities for using machine learning to advance our understanding of biology and improve human health. In summary, machine learning has become an indispensable tool in bioinformatics, enabling researchers to analyze and extract valuable insights from biological data. By deploying machine learning algorithms using command-line tools and techniques such as scikit-learn, researchers can uncover patterns, make predictions, and derive biological knowledge from diverse biological datasets. Machine learning has broad applications in gene expression analysis, sequence analysis, structural biology, and clinical decision support, and it continues to drive innovations in bioinformatics research and biomedical applications. As machine learning techniques continue to evolve and mature, they will play an increasingly important role in advancing our understanding of biological systems and improving human health. Supervised and unsupervised learning are two primary approaches in machine learning that play crucial roles in bioinformatics, aiding in the analysis and interpretation of complex biological data. Next, we delve into the principles of supervised and unsupervised learning, their applications in bioinformatics, and how researchers can deploy these algorithms using command-line tools and techniques.

Understanding Supervised Learning

Supervised learning is a machine learning paradigm where the model is trained on labeled data, meaning that the input data is accompanied by corresponding output labels. The goal of supervised learning is to learn a mapping from inputs to outputs, enabling the model to make predictions or decisions on unseen data. Common supervised learning tasks include classification, regression, and ranking.

In bioinformatics, supervised learning algorithms are applied to various tasks such as predicting gene functions, classifying disease subtypes, and identifying biomarkers associated with clinical outcomes. These algorithms learn from labeled datasets containing features extracted from biological samples and their corresponding labels, allowing researchers to build predictive models that can generalize to unseen data.

Deploying Supervised Learning Algorithms

One widely used command-line tool for supervised learning in bioinformatics is the scikit-learn library in

Python. Researchers can use scikit-learn to train and evaluate supervised learning models, perform feature selection and preprocessing, and tune hyperparameters for optimal performance. For instance, to train a random forest classifier to predict the functional class of genes based on their expression profiles, researchers can use the following command-line script:

bashCopy code

```
python -m pip install scikit-learn python -c "from sklearn.ensemble import RandomForestClassifier; from sklearn.model_selection import train_test_split; from sklearn.metrics import accuracy_score; import numpy as np; data = np.loadtxt('gene_expression_data.txt'); X = data[:, :-1]; y = data[:, -1]; X_train, X_test, y_train, y_test = train_test_split(X, y, test_size=0.2, random_state=42); clf = RandomForestClassifier(); clf.fit(X_train, y_train); y_pred = clf.predict(X_test); accuracy = accuracy_score(y_test, y_pred); print('Accuracy:', accuracy);"
```

This command installs scikit-learn using pip, reads gene expression data from a text file, splits the data into training and testing sets, trains a random forest classifier on the training data, and evaluates its performance on the testing data using accuracy as the evaluation metric.

Understanding Unsupervised Learning

Unsupervised learning, on the other hand, involves training models on unlabeled data, where the objective is to discover underlying patterns, structures, or relationships within the data without explicit guidance. Common unsupervised learning tasks include clustering, dimensionality reduction, and anomaly detection. In bioinformatics, unsupervised learning algorithms are used to uncover hidden structures in biological datasets, identify groups of similar samples, and reduce the dimensionality of high-dimensional data for visualization and exploration.

Deploying Unsupervised Learning Algorithms

One popular command-line tool for unsupervised learning in bioinformatics is the k-means clustering algorithm, which is implemented in various programming languages, including Python and R. Researchers can use k-means clustering to partition biological samples into distinct clusters based on their feature representations.

For example, to perform k-means clustering on gene expression data to identify clusters of co-expressed genes, researchers can use the following command-line script in R:

bashCopy code

```
Rscript -e "data <- read.table('gene_expression_data.txt', header=TRUE); k <- 3; result <- kmeans(data[, -1], centers=k); print(result$cluster);"
```

This command reads gene expression data from a text file, performs k-means clustering on the expression profiles of genes, and assigns each gene to one of the k clusters based on their similarity.

Applications in Bioinformatics

Supervised and unsupervised learning algorithms have broad applications in bioinformatics, including:

Classification of Biological Samples: Supervised learning algorithms are used to classify disease subtypes, predict patient outcomes, and identify biomarkers associated with clinical phenotypes.

Clustering of Gene Expression Data: Unsupervised learning algorithms such as k-means clustering are used to identify groups of co-expressed genes, uncovering potential regulatory pathways and functional modules.

Dimensionality Reduction for Visualization: Unsupervised learning techniques like principal component analysis (PCA) and t-distributed stochastic neighbor embedding (t-SNE) are used to reduce the dimensionality of high-dimensional biological data for visualization and exploration. In summary, supervised and unsupervised learning algorithms are indispensable tools in bioinformatics, enabling researchers to analyze, interpret, and extract valuable insights from complex biological datasets. By deploying these algorithms using command-line tools and techniques, researchers can build predictive models, uncover hidden structures, and gain deeper insights into the underlying biology of organisms and diseases. Supervised learning algorithms are used for tasks such as classification and regression, while unsupervised learning algorithms are employed for clustering, dimensionality reduction, and anomaly detection. Together, these approaches empower bioinformaticians to address a wide range of biological questions and accelerate discoveries in the field of life sciences.

Chapter 6: Predictive Modeling and Classification

Predictive modeling techniques play a pivotal role in bioinformatics, allowing researchers to build models that can predict biological outcomes, classify samples, and identify patterns in complex datasets. Next, we explore various predictive modeling techniques used in bioinformatics, their applications, and how researchers can deploy these techniques using command-line tools and techniques.

Understanding Predictive Modeling

Predictive modeling involves the development of mathematical models that can make predictions based on input data. These models learn from historical data to uncover patterns and relationships, which are then used to make predictions on new, unseen data. Predictive modeling encompasses a wide range of techniques, including regression, classification, and ensemble methods.

In bioinformatics, predictive modeling is applied to various biological datasets, such as gene expression profiles, DNA sequences, and clinical data, to predict phenotypic outcomes, classify disease subtypes, and identify biomarkers associated with diseases.

Deploying Predictive Modeling Techniques

One commonly used command-line tool for predictive modeling in bioinformatics is the scikit-learn library in Python. Researchers can use scikit-learn to train and evaluate predictive models, perform feature selection and preprocessing, and tune hyperparameters for optimal performance.

For example, to train a logistic regression model to predict the risk of a disease based on genetic variants, researchers can use the following command-line script: bashCopy code

```
python -m pip install scikit-learn python -c "from sklearn.linear_model import LogisticRegression; from sklearn.model_selection import train_test_split; from sklearn.metrics import accuracy_score; import numpy as np; data = np.loadtxt('genetic_variant_data.txt'); X = data[:, :-1]; y = data[:, -1]; X_train, X_test, y_train, y_test = train_test_split(X, y, test_size=0.2, random_state=42); clf = LogisticRegression(); clf.fit(X_train, y_train); y_pred = clf.predict(X_test); accuracy = accuracy_score(y_test, y_pred); print('Accuracy:', accuracy);"
```

This command installs scikit-learn using pip, reads genetic variant data from a text file, splits the data into training and testing sets, trains a logistic regression model on the training data, and evaluates its performance on the testing data using accuracy as the evaluation metric.

Applications of Predictive Modeling in Bioinformatics

Predictive modeling techniques have diverse applications in bioinformatics, including:

Disease Prediction: Predictive models are used to predict the risk of developing diseases based on genetic, environmental, and clinical factors, aiding in early diagnosis and intervention.

Drug Response Prediction: Models can predict the efficacy and toxicity of drugs based on genomic and pharmacological data, enabling personalized treatment strategies.

Biomarker Identification: Predictive models are used to identify biomarkers associated with diseases, drug response, and treatment outcomes, facilitating targeted therapies and precision medicine.

Functional Annotation: Models predict the functions of genes, proteins, and non-coding RNAs based on their sequence, structure, and expression profiles, aiding in functional annotation and interpretation of genomic data.

Challenges and Future Directions

Despite their utility, predictive modeling techniques face several challenges in bioinformatics, including data heterogeneity, model interpretability, and ethical considerations related to data privacy and bias. However, ongoing advancements in machine learning algorithms, data integration techniques, and computational resources hold promise for addressing these challenges and unlocking new opportunities for predictive modeling in bioinformatics.

In summary, predictive modeling techniques are indispensable tools in bioinformatics, enabling researchers to make predictions, classify samples, and identify patterns in complex biological datasets. By deploying these techniques using command-line tools and techniques such as scikit-learn, researchers can build predictive models that aid in disease prediction, drug response prediction, biomarker identification, and functional annotation. Despite facing challenges, predictive modeling continues to drive innovations in bioinformatics research and holds tremendous potential for advancing our understanding of biological systems and improving human health.

Classification Algorithms and Applications in Bioinformatics

Classification algorithms are fundamental tools in bioinformatics, facilitating the categorization of biological data into discrete classes or categories based on their features. Next, we explore various classification algorithms commonly used in bioinformatics, their applications, and how researchers can deploy these techniques using command-line tools and techniques.

Understanding Classification Algorithms

Classification algorithms are supervised learning techniques that assign predefined labels or categories to input data based on their features. These algorithms learn from labeled training data to build models that can predict the class labels of unseen instances. Common classification algorithms include decision trees, support vector machines (SVM), random forests, and neural networks.

In bioinformatics, classification algorithms are applied to various tasks such as disease diagnosis, protein function prediction, and drug target identification. These algorithms analyze biological data, such as gene expression profiles, DNA sequences, and protein structures, to classify samples into different phenotypic or functional categories.

Deploying Classification Algorithms

One popular command-line tool for classification in bioinformatics is the scikit-learn library in Python. Researchers can use scikit-learn to train and evaluate classification models, perform feature selection and preprocessing, and tune hyperparameters for optimal performance.

For example, to train a support vector machine (SVM) classifier to predict the subcellular localization of proteins based on their amino acid sequences, researchers can use the following command-line script:

bashCopy code

```
python -m pip install scikit-learn python -c "from sklearn.svm import SVC; from sklearn.model_selection import train_test_split; from sklearn.metrics import accuracy_score; import numpy as np; data = np.loadtxt('protein_sequence_data.txt'); X = data[:, :-1]; y = data[:, -1]; X_train, X_test, y_train, y_test = train_test_split(X, y, test_size=0.2, random_state=42); clf = SVC(); clf.fit(X_train, y_train); y_pred = clf.predict(X_test); accuracy = accuracy_score(y_test, y_pred); print('Accuracy:', accuracy);"
```

This command installs scikit-learn using pip, reads protein sequence data from a text file, splits the data into training and testing sets, trains an SVM classifier on the training data, and evaluates its performance on the testing data using accuracy as the evaluation metric.

Applications of Classification in Bioinformatics

Classification algorithms have diverse applications in bioinformatics, including:

Disease Diagnosis: Classification models predict disease outcomes based on clinical and genomic data, aiding in the early diagnosis and prognosis of diseases.

Protein Function Prediction: Models classify proteins into functional categories based on sequence, structure, and interaction data, facilitating the annotation of protein functions and pathways.

Drug Target Identification: Classification algorithms identify potential drug targets by classifying proteins into druggable and non-druggable categories, guiding drug discovery and development efforts.

Species Classification: Models classify organisms into different taxonomic groups based on their genetic sequences, enabling the identification and characterization of species in environmental and metagenomic samples.

Challenges and Future Directions

Despite their effectiveness, classification algorithms face several challenges in bioinformatics, including data heterogeneity, class imbalance, and interpretability of models. Future research directions include the development of interpretable and robust classification algorithms, integration of multi-omics data for improved predictive performance, and ethical considerations related to data privacy and bias.

In summary, classification algorithms are essential tools in bioinformatics, enabling researchers to categorize biological data and extract meaningful insights from complex datasets. By deploying these algorithms using command-line tools and techniques such as scikit-learn, researchers can build classification models that aid in disease diagnosis, protein function prediction, drug target identification, and species classification. Despite facing challenges, classification algorithms continue to drive innovations in bioinformatics research and hold promise for advancing our understanding of biological systems and improving human health.

Chapter 7: Clustering and Dimensionality Reduction Techniques

Clustering methods are essential tools in bioinformatics, enabling researchers to uncover patterns, group similar entities, and identify hidden structures within biological datasets. Next, we delve into various clustering algorithms commonly used in bioinformatics, their applications, and how researchers can deploy these techniques using command-line tools and techniques.

Understanding Clustering Methods

Clustering methods are unsupervised learning techniques that partition data into groups, or clusters, based on the similarity of their features. These algorithms aim to maximize intra-cluster similarity while minimizing inter-cluster similarity, grouping together entities with similar characteristics.

Common clustering algorithms include k-means, hierarchical clustering, and density-based clustering (e.g., DBSCAN). These algorithms differ in their approach to defining clusters and handling data with different structures and distributions.

In bioinformatics, clustering methods are applied to diverse biological datasets, such as gene expression profiles, protein interaction networks, and metagenomic samples, to identify co-expressed genes, protein complexes, and microbial communities.

Deploying Clustering Methods

One widely used command-line tool for clustering analysis in bioinformatics is the scikit-learn library in

Python. Researchers can leverage scikit-learn to implement various clustering algorithms, perform clustering on biological datasets, and evaluate clustering results.

For instance, to perform k-means clustering on gene expression data to identify co-expressed gene modules, researchers can use the following command-line script:

bashCopy code

```
python -m pip install scikit-learn python -c "from sklearn.cluster import KMeans; from sklearn.preprocessing import StandardScaler; import numpy as np; gene_expression_data = np.loadtxt('gene_expression_data.txt'); scaled_data = StandardScaler().fit_transform(gene_expression_data); kmeans = KMeans(n_clusters=5, random_state=42); clusters = kmeans.fit_predict(scaled_data); print(clusters);"
```

This command installs scikit-learn using pip, reads gene expression data from a text file, scales the data to zero mean and unit variance, performs k-means clustering with five clusters, and assigns each gene to a cluster.

Applications of Clustering in Bioinformatics

Clustering methods find numerous applications in bioinformatics, including:

Gene Expression Analysis: Clustering is used to identify co-expressed genes and infer gene regulatory networks from transcriptomic data.

Protein Interaction Analysis: Clustering helps identify protein complexes and functional modules in protein-protein interaction networks, aiding in the understanding of cellular processes.

Metagenomic Community Analysis: Clustering is applied to metagenomic data to classify microbial communities, identify taxonomic groups, and assess microbial diversity and composition.

Drug Discovery: Clustering methods group compounds with similar chemical structures or biological activities, facilitating the identification of lead compounds and drug repurposing.

Challenges and Future Directions

Despite their versatility, clustering methods face challenges in handling high-dimensional, noisy, and heterogeneous biological data. Future research directions include the development of robust clustering algorithms capable of handling large-scale omics data, integration of multi-omics data for improved clustering accuracy, and incorporation of domain knowledge into clustering analysis for better interpretation of results.

In summary, clustering methods are indispensable tools in bioinformatics, enabling researchers to uncover patterns and structures within biological datasets. By deploying these techniques using command-line tools and libraries such as scikit-learn, researchers can perform clustering analysis on diverse biological datasets, including gene expression data, protein interaction networks, and metagenomic samples. Despite facing challenges, clustering methods continue to drive innovations in bioinformatics research and hold promise for advancing our understanding of biological systems and addressing complex biomedical challenges.

Dimensionality Reduction Techniques in Bioinformatics

Dimensionality reduction techniques are essential tools in bioinformatics, enabling researchers to extract meaningful information from high-dimensional biological datasets while reducing computational complexity and addressing issues such as overfitting and curse of dimensionality. Next, we explore various dimensionality reduction techniques commonly used in bioinformatics, their applications, and how researchers can deploy these techniques using command-line tools and techniques.

Understanding Dimensionality Reduction

Dimensionality reduction techniques aim to reduce the number of features or variables in a dataset while preserving its essential characteristics. These techniques transform high-dimensional data into a lower-dimensional representation, making it more manageable for analysis and visualization without losing critical information.

Common dimensionality reduction techniques include principal component analysis (PCA), t-distributed stochastic neighbor embedding (t-SNE), and uniform manifold approximation and projection (UMAP). These techniques differ in their approach to capturing the underlying structure of the data and preserving its intrinsic properties.

In bioinformatics, dimensionality reduction techniques are applied to various types of biological data, such as gene expression profiles, single-cell RNA sequencing data, and imaging data, to uncover hidden patterns, identify key features, and visualize complex datasets.

Deploying Dimensionality Reduction Techniques

One widely used command-line tool for dimensionality reduction in bioinformatics is the scikit-learn library in Python. Researchers can utilize scikit-learn to implement various dimensionality reduction algorithms, apply them to biological datasets, and visualize the reduced-dimensional representations.

For example, to perform principal component analysis (PCA) on gene expression data to reduce its dimensionality and visualize the principal components, researchers can use the following command-line script:

bashCopy code

```
python -m pip install scikit-learn python -c "from sklearn.decomposition import PCA; from sklearn.preprocessing import StandardScaler; import numpy as np; gene_expression_data = np.loadtxt('gene_expression_data.txt'); scaled_data = StandardScaler().fit_transform(gene_expression_data); pca = PCA(n_components=2); pca_data = pca.fit_transform(scaled_data); print(pca_data);"
```

This command installs scikit-learn using pip, reads gene expression data from a text file, scales the data to zero mean and unit variance, performs PCA to reduce the dimensionality to two components, and prints the reduced-dimensional representation.

Applications of Dimensionality Reduction in Bioinformatics

Dimensionality reduction techniques find numerous applications in bioinformatics, including:

Visualization of High-Dimensional Data: Dimensionality reduction techniques enable the visualization of high-dimensional biological datasets in two or three dimensions, facilitating the exploration and interpretation of complex data structures.

Feature Selection and Compression: Dimensionality reduction helps identify key features or variables contributing to the variability in the data and compresses the data representation while retaining its essential characteristics.

Integration of Multi-Omics Data: Dimensionality reduction techniques can integrate multiple omics datasets (e.g., genomics, transcriptomics, proteomics) into a lower-dimensional space, enabling the analysis of integrated omics data and identification of biomarkers and disease signatures.

Single-Cell Data Analysis: Dimensionality reduction is applied to single-cell RNA sequencing data to visualize cellular heterogeneity, identify cell types, and uncover gene regulatory networks at the single-cell level.

Challenges and Future Directions

Despite their utility, dimensionality reduction techniques face challenges in handling nonlinear relationships, preserving local structures, and interpreting reduced-dimensional representations. Future research directions include the development of nonlinear dimensionality reduction methods, integration of domain knowledge into dimensionality reduction analysis, and exploration of deep learning approaches for dimensionality reduction in bioinformatics.

In summary, dimensionality reduction techniques play a crucial role in bioinformatics, enabling researchers to analyze and visualize high-dimensional biological datasets effectively. By deploying these techniques using command-line tools and libraries such as scikit-learn, researchers can uncover hidden patterns, identify key features, and gain insights into complex biological systems. Despite facing challenges, dimensionality reduction techniques continue to drive innovations in bioinformatics research and hold promise for advancing our understanding of biological processes and addressing biomedical challenges.

Chapter 8: Deep Learning Applications in Bioinformatics

Deep learning has emerged as a powerful tool in bioinformatics, revolutionizing the analysis and interpretation of biological data. Next, we explore the fundamentals of deep learning, its applications in bioinformatics, and how researchers can deploy deep learning techniques using command-line tools and frameworks.

Understanding Deep Learning

Deep learning is a subfield of machine learning that utilizes neural networks with multiple layers (hence the term "deep") to automatically learn hierarchical representations of data. These neural networks are trained using large datasets to perform tasks such as classification, regression, clustering, and feature extraction.

Deep learning architectures commonly used in bioinformatics include convolutional neural networks (CNNs) for image analysis, recurrent neural networks (RNNs) for sequence analysis, and deep belief networks (DBNs) for unsupervised learning tasks.

Deploying Deep Learning Techniques

One widely used command-line tool for deep learning in bioinformatics is TensorFlow, an open-source machine learning framework developed by Google. Researchers can leverage TensorFlow to implement various deep learning architectures, train neural networks on biological datasets, and perform tasks such as gene expression analysis, protein structure prediction, and drug discovery.

For example, to train a convolutional neural network (CNN) on histopathology images for cancer classification, researchers can use the following command-line script:

bashCopy code

```
pip install tensorflow python train_cnn.py --data_dir=/path/to/histopathology_images --num_classes=2 --epochs=50
```

This command installs TensorFlow using pip, specifies the directory containing histopathology images and the number of classes (e.g., cancerous vs. non-cancerous), and trains the CNN model for 50 epochs.

Applications of Deep Learning in Bioinformatics

Deep learning has numerous applications in bioinformatics, including:

Sequence Analysis: Deep learning techniques such as recurrent neural networks (RNNs) and long short-term memory (LSTM) networks are used for DNA sequence analysis, protein sequence prediction, and RNA structure prediction.

Image Analysis: Convolutional neural networks (CNNs) are applied to analyze biological images, such as histopathology slides, microscopy images, and cellular imaging data, for tasks such as cell classification, tumor detection, and image segmentation.

Drug Discovery: Deep learning models are used to predict molecular properties, identify potential drug candidates, and analyze drug-target interactions, accelerating the drug discovery process.

Functional Genomics: Deep learning techniques enable the prediction of gene functions, regulatory elements, and protein-protein interactions from genomic and proteomic data, aiding in the interpretation of biological pathways and networks.

Challenges and Future Directions

Despite their effectiveness, deep learning techniques face challenges in handling small datasets, interpretability of models, and generalization to unseen data. Future research directions include the development of explainable deep learning models, integration of multi-omics data for comprehensive analysis, and exploration of deep reinforcement learning for drug discovery and personalized medicine.

In summary, deep learning has transformed the field of bioinformatics, enabling researchers to analyze complex biological data and extract valuable insights. By deploying deep learning techniques using command-line tools and frameworks such as TensorFlow, researchers can tackle diverse bioinformatics tasks, from sequence analysis to image classification and drug discovery. Despite facing challenges, deep learning continues to drive innovations in bioinformatics research and holds promise for advancing our understanding of biological systems and addressing critical biomedical challenges.

Deep Learning Models and Applications for Biological Data

Deep learning has emerged as a transformative approach for analyzing and interpreting various types of biological data, ranging from genomic sequences to medical images. Next, we delve into the diverse deep learning models utilized in bioinformatics and explore their applications across different domains. Additionally, we provide insights into how researchers can deploy these models using command-line tools and frameworks.

Understanding Deep Learning Models

Deep learning models are neural network architectures consisting of multiple layers, enabling them to automatically learn hierarchical representations from raw data. These models have revolutionized the field of bioinformatics by providing powerful tools for tasks such as sequence analysis, image classification, and drug discovery.

Common deep learning architectures used in bioinformatics include convolutional neural networks (CNNs) for image analysis, recurrent neural networks (RNNs) for sequential data, and generative adversarial networks (GANs) for data generation. Each of these architectures is tailored to specific types of biological data and tasks.

Deploying Deep Learning Models

Deploying deep learning models in bioinformatics often involves using specialized frameworks such as TensorFlow, Keras, or PyTorch. These frameworks provide command-line interfaces for training and deploying models, along with APIs for data preprocessing and evaluation.

For example, to train a CNN model for image classification using TensorFlow, researchers can use the following command-line script:

bashCopy code

```
pip install tensorflow python train_cnn.py --data_dir=/path/to/image_data --num_classes=2 --epochs=50
```

This command installs TensorFlow using pip, specifies the directory containing image data and the number of classes, and trains the CNN model for 50 epochs.

Applications of Deep Learning in Bioinformatics

Deep learning models find extensive applications across various domains of bioinformatics, including:

Genomic Sequence Analysis: RNNs and attention-based models are used for DNA sequence classification, motif discovery, and gene expression prediction.

Protein Structure Prediction: CNNs and graph neural networks (GNNs) are employed for predicting protein structures, identifying functional residues, and analyzing protein-protein interactions.

Medical Image Analysis: CNNs are applied to analyze medical images such as MRI scans, CT scans, and histopathology slides for tasks like tumor detection, organ segmentation, and disease diagnosis.

Drug Discovery: GANs and reinforcement learning models are utilized for de novo drug design, virtual screening, and predicting drug-target interactions.

Challenges and Future Directions

Despite their effectiveness, deep learning models in bioinformatics face challenges such as data scarcity, interpretability, and generalization to unseen data. Future research directions include the development of explainable deep learning models, integration of multi-omics data for comprehensive analysis, and exploration of reinforcement learning for personalized medicine.

In summary, deep learning models have revolutionized bioinformatics by providing powerful tools for analyzing and interpreting biological data. By deploying these models using command-line tools and frameworks, researchers can tackle diverse bioinformatics tasks, from sequence analysis to image classification and drug discovery. Despite facing challenges, deep learning continues to drive innovations in bioinformatics research and holds promise for advancing our understanding of biological systems and addressing critical biomedical challenges.

Chapter 9: Network Analysis and Graph Theory in Bioinformatics

Network Analysis Fundamentals

Network analysis is a powerful technique used in various fields, including bioinformatics, to study complex relationships and interactions among biological entities. Next, we explore the fundamentals of network analysis, its applications in bioinformatics, and how researchers can deploy network analysis techniques using command-line tools and software packages.

Understanding Networks

A network, also known as a graph, consists of nodes (vertices) connected by edges (links). Nodes represent entities, such as genes, proteins, or metabolites, while edges represent relationships or interactions between these entities. Networks can be classified based on their properties, such as directed vs. undirected, weighted vs. unweighted, and static vs. dynamic.

Deploying Network Analysis Techniques

Various command-line tools and software packages are available for performing network analysis in bioinformatics. One widely used tool is Cytoscape, an open-source platform for visualizing and analyzing biological networks. To analyze a protein-protein interaction network using Cytoscape, researchers can follow these steps:

Data Preparation: Prepare the protein-protein interaction data in a supported format, such as a tab-separated values (TSV) file or a network file format (SIF).

Installation: Install Cytoscape on your computer using the following command:

bashCopy code

brew install cytoscape

This command installs Cytoscape on macOS using Homebrew. Alternatively, you can download and install Cytoscape from the official website for other operating systems.

Network Import: Import the protein-protein interaction data into Cytoscape using the File > Import > Network option.

Visualization: Visualize the network using various layout algorithms and styles available in Cytoscape to gain insights into the interactions between proteins.

Applications of Network Analysis in Bioinformatics

Network analysis finds diverse applications in bioinformatics, including:

Gene Regulatory Networks: Studying interactions between genes to understand gene regulation mechanisms and identify key regulatory elements.

Protein-Protein Interaction Networks: Analyzing protein-protein interaction networks to identify protein complexes, pathways, and disease-associated proteins.

Metabolic Networks: Modeling metabolic pathways and analyzing metabolic networks to predict metabolic fluxes, identify drug targets, and understand metabolic diseases.

Signaling Networks: Investigating signaling pathways and signal transduction networks to elucidate cellular processes and identify potential drug targets for cancer therapy.

Challenges and Future Directions

Despite its utility, network analysis in bioinformatics faces challenges such as data integration, network inference, and scalability. Future research directions include the development of advanced algorithms for network analysis, integration of multi-omics data for comprehensive network modeling, and exploration of dynamic network analysis techniques to capture temporal changes in biological systems. In summary, network analysis is a valuable technique in bioinformatics for studying complex biological systems and interactions. By deploying network analysis techniques using command-line tools and software packages like Cytoscape, researchers can gain insights into the structure, function, and dynamics of biological networks. Despite facing challenges, network analysis continues to drive discoveries in bioinformatics research and holds promise for understanding the complexity of living organisms and addressing biomedical challenges.

Graph Theory Applications in Biological Networks

Graph theory provides a powerful framework for analyzing and modeling complex relationships in biological systems. Next, we explore the applications of graph theory in understanding biological networks, including protein-protein interaction networks, gene regulatory networks, metabolic networks, and signaling networks. We also discuss how researchers can deploy graph theory techniques using command-line tools and software packages.

Understanding Biological Networks as Graphs

Biological networks can be represented as graphs, where nodes represent biological entities such as genes, proteins, metabolites, or cellular components, and edges represent relationships or interactions between these entities. Graph theory provides a rich set of mathematical tools and algorithms for analyzing the structure, function, and dynamics of biological networks.

Deploying Graph Theory Techniques

Several command-line tools and software packages are available for deploying graph theory techniques in bioinformatics. One widely used tool is NetworkX, a Python library for the creation, manipulation, and analysis of complex networks. To analyze a protein-protein interaction network using NetworkX, researchers can follow these steps:

Installation: Install NetworkX on your computer using the following command:

bashCopy code

```
pip install networkx
```

This command installs NetworkX and its dependencies on your Python environment.

Data Preparation: Prepare the protein-protein interaction data in a supported format, such as an edge list or an adjacency matrix.

Network Construction: Construct the protein-protein interaction network using NetworkX by loading the data into a graph object:

```python
pythonCopy code
import networkx as nx # Load protein-protein
interaction data G =
nx.read_edgelist('protein_interactions.txt')
```

This Python code reads the protein-protein interaction data from a text file and constructs a graph object **G** representing the network.

Network Analysis: Analyze the protein-protein interaction network using various graph theory algorithms provided by NetworkX, such as degree centrality, betweenness centrality, clustering coefficient, and shortest path length.

```python
pythonCopy code
# Calculate degree centrality degree_centrality =
nx.degree_centrality(G)
```

This code calculates the degree centrality of nodes in the network, which measures the importance of nodes based on the number of edges incident to them.

Applications of Graph Theory in Biological Networks

Graph theory finds diverse applications in understanding biological networks:

Protein-Protein Interaction Networks: Analyzing the topology of protein-protein interaction networks to identify hub proteins, network motifs, and modules associated with specific biological functions or diseases.

Gene Regulatory Networks: Modeling the interactions between genes and transcription factors to elucidate gene regulatory mechanisms, identify regulatory motifs, and predict gene expression patterns.

Metabolic Networks: Modeling metabolic pathways and analyzing metabolic networks to understand metabolic fluxes, identify key metabolic enzymes, and predict metabolic phenotypes.

Signaling Networks: Analyzing signal transduction pathways and signaling networks to unravel cellular signaling cascades, identify key signaling molecules, and elucidate their roles in cellular processes and diseases.

Challenges and Future Directions

Despite its utility, graph theory analysis in bioinformatics faces challenges such as network inference, data integration, and scalability. Future research directions include the development of advanced graph theory algorithms for analyzing large-scale biological networks, integration of multi-omics data for comprehensive network modeling, and exploration of dynamic network analysis techniques to capture temporal changes in biological systems. In summary, graph theory provides a powerful framework for analyzing and modeling complex biological networks. By deploying graph theory techniques using command-line tools and software packages like NetworkX, researchers can gain insights into the structure, function, and dynamics of biological networks across different domains. Despite facing challenges, graph theory continues to drive discoveries in bioinformatics research and holds promise for understanding the complexity of living organisms and addressing biomedical challenges.

Chapter 10: Big Data Management and Tools for Bioinformatics

In the era of big data, where the volume, velocity, and variety of data are constantly increasing, managing and extracting insights from large datasets pose significant challenges in various domains, including bioinformatics. Next, we explore the challenges associated with big data management and discuss potential solutions and techniques to address them. We also provide insights into how researchers can deploy these techniques using command-line tools and software packages.

Understanding Big Data Management

Big data management involves storing, processing, and analyzing massive volumes of data efficiently and effectively. The challenges in big data management stem from the sheer scale of data, as well as its complexity, heterogeneity, and dynamic nature. Traditional data management systems and techniques are often inadequate to handle big data due to limitations in storage capacity, processing speed, and scalability.

Deploying Distributed File Systems

One approach to address the challenges of big data management is to deploy distributed file systems that distribute data across multiple nodes in a cluster for parallel processing. Apache Hadoop is a popular open-source framework for distributed storage and processing of big data. To deploy a Hadoop cluster, researchers can follow these steps:

Installation: Install Apache Hadoop on each node of the cluster using the following command:

```
bashCopy code
wget
```
https://archive.apache.org/dist/hadoop/core/hadoop-3.3.1/hadoop-3.3.1.tar.gz tar -xzvf hadoop-3.3.1.tar.gz

This command downloads the Hadoop distribution and extracts it into a directory on each node.

Configuration: Configure Hadoop by editing the **hadoop-env.sh**, **core-site.xml**, **hdfs-site.xml**, and **mapred-site.xml** files to specify cluster settings such as the number of nodes, memory allocation, and data directories.

Cluster Initialization: Initialize the Hadoop cluster using the following command:

```
bashCopy code
hdfs namenode -format
```

This command formats the Hadoop distributed file system (HDFS) on the Namenode, preparing it for use.

Cluster Startup: Start the Hadoop services on each node of the cluster using the following command:

```
bashCopy code
start-all.sh
```

This command starts the Hadoop daemons, including the Namenode, Datanode, ResourceManager, and NodeManager, on all nodes of the cluster.

Data Partitioning and Sharding

Another challenge in big data management is efficiently partitioning and sharding data to distribute it across multiple nodes for parallel processing. Partitioning involves dividing large datasets into smaller chunks based on certain criteria, such as key range or hash value, while sharding involves distributing these partitions across different nodes in the cluster. MongoDB, a NoSQL database, provides built-in support for sharding, allowing researchers to distribute data across multiple shards for horizontal scalability.

Data Compression and Encoding

Data compression and encoding techniques play a crucial role in reducing storage and bandwidth requirements for big data management. Compression algorithms such as gzip, bzip2, and Snappy can significantly reduce the size of large datasets without loss of information. Researchers can deploy these compression algorithms using command-line tools or libraries available in programming languages like Python or Java.

Distributed Computing Paradigms

In addition to distributed file systems, distributed computing paradigms such as MapReduce and Apache Spark offer scalable and fault-tolerant solutions for processing big data. MapReduce divides a computation into smaller tasks and distributes them across multiple nodes in a cluster, while Spark provides in-memory processing capabilities for faster data processing. Researchers can deploy MapReduce or Spark jobs using command-line tools or APIs provided by these frameworks.

In summary, big data management presents various challenges, including storage, processing, and analysis of massive volumes of data. By deploying distributed file systems, data partitioning and sharding techniques, data compression and encoding methods, and distributed computing paradigms, researchers can overcome these challenges and efficiently manage big data in bioinformatics and other domains. Command-line tools and software packages provide convenient means to deploy these techniques and harness the power of big data for scientific discovery and innovation.

Tools and Platforms for Big Data Analysis in Bioinformatics

In the field of bioinformatics, the explosion of biological data has necessitated the development of specialized tools and platforms for analyzing large-scale datasets efficiently. Next, we explore various tools and platforms tailored for big data analysis in bioinformatics. We discuss their features, capabilities, and deployment methods, including command-line interface (CLI) commands, to empower researchers in their data analysis endeavors.

Hadoop Ecosystem

Apache Hadoop, an open-source framework for distributed storage and processing of big data, forms the backbone of many bioinformatics data analysis platforms. The Hadoop ecosystem comprises several complementary tools and frameworks, each serving specific purposes in the big data analysis pipeline.

Hadoop Distributed File System (HDFS): HDFS is a distributed file system that stores large datasets across multiple nodes in a Hadoop cluster. Researchers can deploy HDFS using CLI commands to store and manage bioinformatics data efficiently.

bashCopy code

```
hdfs dfs -mkdir /user/input
```

This command creates a directory named **/user/input** in the HDFS filesystem.

MapReduce: MapReduce is a programming model and processing engine for parallel processing of large datasets. Researchers can write MapReduce programs in Java,

Python, or other languages to perform distributed computation on bioinformatics data stored in HDFS.

bashCopy code

```
hadoop jar <path-to-jar> <main-class> <input-path> <output-path>
```

This command submits a MapReduce job to the Hadoop cluster for execution.

Apache Spark: Apache Spark is a fast and general-purpose cluster computing system that provides in-memory data processing capabilities. Researchers can deploy Spark to perform iterative and interactive data analysis tasks on large-scale bioinformatics datasets.

bashCopy code

```
spark-submit --class <main-class> --master <master-url> <path-to-jar> <input-path> <output-path>
```

This command submits a Spark job to the cluster for execution.

Bioinformatics-specific Platforms

Several bioinformatics-specific platforms leverage the capabilities of the Hadoop ecosystem to provide specialized tools and workflows for analyzing biological data. These platforms offer pre-built analytics pipelines, scalable infrastructure, and user-friendly interfaces tailored to the needs of bioinformatics researchers.

Galaxy: Galaxy is an open-source, web-based platform for data-intensive biomedical research. It provides a user-friendly interface for designing, executing, and sharing bioinformatics workflows. Researchers can deploy Galaxy using Docker containers or virtual machines.

bashCopy code

```
docker run -d -p 8080:80 quay.io/galaxy/docker:20.05
```

This command pulls the Galaxy Docker image from the Docker registry and starts a Galaxy server on port 8080.

SeqWare: SeqWare is a distributed computing platform for genomics data analysis. It offers a suite of bioinformatics tools and workflows for processing next-generation sequencing (NGS) data at scale. Researchers can deploy SeqWare using Docker or Kubernetes.

bashCopy code

kubectl create -f seqware.yaml

This command deploys the SeqWare platform on a Kubernetes cluster using the configuration specified in the **seqware.yaml** file.

Cloud-based Solutions

Cloud computing platforms such as Amazon Web Services (AWS), Google Cloud Platform (GCP), and Microsoft Azure offer scalable infrastructure and managed services for big data analysis in bioinformatics. Researchers can leverage these platforms to access high-performance computing resources, storage, and analysis tools on-demand.

AWS Batch: AWS Batch is a managed service for running batch computing workloads on AWS. Researchers can use AWS Batch to execute bioinformatics workflows in the cloud, taking advantage of AWS's scalable infrastructure and pricing model.

bashCopy code

aws batch submit-job --job-definition <job-definition> --job-name <job-name> --job-queue <job-queue>

This command submits a batch job to AWS Batch for execution.

GCP Dataflow: Google Cloud Dataflow is a fully managed service for stream and batch processing of data on GCP.

Researchers can use Dataflow to develop and deploy data processing pipelines for bioinformatics analysis tasks.
bashCopy code

```
gcloud dataflow jobs run <job-name> --gcs-location
<template-location> --parameters input=<input-
path>,output=<output-path>
```

This command launches a Dataflow job with the specified input and output paths.

In summary, a variety of tools and platforms are available for conducting big data analysis in bioinformatics. From the Hadoop ecosystem to bioinformatics-specific platforms and cloud-based solutions, researchers have access to a rich set of resources for processing and analyzing large-scale biological datasets. By leveraging these tools and platforms, researchers can accelerate their research, uncover new insights, and advance our understanding of complex biological systems.

BOOK 4
MASTERING BIOSTATISTICS IN BIOINFORMATICS
ADVANCED METHODS AND APPLICATION

ROB BOTWRIGHT

Chapter 1: Introduction to Biostatistics in Bioinformatics

Biostatistics plays a crucial role in the field of bioinformatics, serving as the cornerstone for analyzing and interpreting complex biological data. Next, we delve into the fundamentals of biostatistics, its applications in bioinformatics, and explore various statistical techniques used to extract meaningful insights from biological datasets. We will discuss how biostatistics aids in experimental design, hypothesis testing, data visualization, and modeling, empowering researchers to make informed decisions and draw reliable conclusions from their experiments.

Understanding Biostatistics

Biostatistics is the application of statistical methods to biological data. It encompasses the design, analysis, and interpretation of experiments and observational studies in fields such as genetics, epidemiology, and clinical trials. Biostatisticians develop and apply statistical models and techniques to address research questions, assess the validity of experimental results, and quantify uncertainty.

Applications in Bioinformatics

In bioinformatics, biostatistics plays a central role in analyzing various types of biological data, including genomic sequences, gene expression profiles, protein structures, and metabolic pathways. Biostatistical methods are used to:

Experimental Design: Biostatistics helps researchers design experiments that are statistically sound and capable of providing reliable results. This includes determining sample sizes, randomization procedures, and control groups to minimize bias and maximize the power of the study.

Hypothesis Testing: Biostatistical tests such as t-tests, chi-square tests, and ANOVA are used to assess the significance of observed differences between groups or conditions. These tests help researchers determine whether the results of their experiments are due to true biological effects or random variability.

Data Visualization: Biostatistics provides tools for visualizing biological data through graphs, charts, and heatmaps. Visualization techniques such as scatter plots, box plots, and histograms allow researchers to explore patterns, trends, and relationships in the data, facilitating the interpretation of complex datasets.

Modeling and Prediction: Biostatistical models such as linear regression, logistic regression, and machine learning algorithms are used to build predictive models based on biological data. These models can be used to predict outcomes, classify samples, and identify biomarkers associated with disease or phenotype.

Statistical Techniques in Bioinformatics

Several statistical techniques are commonly used in bioinformatics to analyze biological data:

Descriptive Statistics: Descriptive statistics summarize the characteristics of a dataset using measures such as mean, median, standard deviation, and percentiles. These summary statistics provide insights into the central tendency, variability, and distribution of the data.

Inferential Statistics: Inferential statistics are used to make inferences or predictions about a population based on a sample of data. This includes hypothesis testing, confidence intervals, and estimation of population parameters.

Bayesian Statistics: Bayesian statistics is a probabilistic approach to inference that incorporates prior knowledge and uncertainty into statistical analysis. Bayesian methods are particularly useful in bioinformatics for modeling complex biological systems and integrating multiple sources of data.

Survival Analysis: Survival analysis is used to analyze time-to-event data, such as time until death or disease recurrence. Kaplan-Meier curves, Cox proportional hazards models, and log-rank tests are commonly used techniques in survival analysis.

In summary, biostatistics plays a vital role in bioinformatics by providing the tools and techniques necessary to analyze and interpret biological data. From experimental design to hypothesis testing, data visualization, and modeling, biostatistics underpins every aspect of bioinformatics research. By applying rigorous statistical methods, researchers can derive meaningful insights from biological datasets, advance scientific knowledge, and contribute to discoveries in fields such as genomics, proteomics, and personalized medicine.

Statistical methods form the backbone of biomedical data analysis, enabling researchers to extract meaningful insights, identify patterns, and draw reliable conclusions from complex datasets. Next, we explore the importance of statistical methods in biomedical research, highlighting their role in experimental design, hypothesis testing, data interpretation, and decision-making.

Foundation of Biomedical Data Analysis

Biomedical research generates vast amounts of data from various sources, including clinical trials, genomic studies, imaging modalities, and electronic health records. Statistical methods provide the framework for organizing, analyzing, and interpreting this data, facilitating evidence-based decision-making in healthcare and biomedical research.

Experimental Design

One of the primary roles of statistical methods in biomedical research is in experimental design. Well-designed experiments are essential for generating valid and reliable results. Statistical principles guide researchers in determining sample sizes, allocating resources, and implementing randomization and blinding techniques to minimize bias and maximize the power of the study. For example, in clinical trials, randomization ensures that treatment groups are comparable, while blinding reduces the risk of bias in outcome assessment.

Hypothesis Testing

Statistical hypothesis testing is a fundamental tool for assessing the significance of observed differences or associations in biomedical data. Whether comparing treatment outcomes, evaluating the association between genetic variants and disease risk, or testing the efficacy of diagnostic tests, hypothesis testing allows researchers to determine whether observed effects are statistically significant or due to chance. Common statistical tests include t-tests, chi-square tests, analysis of variance (ANOVA), and regression analysis.

Data Interpretation and Inference

Statistical methods provide the framework for interpreting and drawing conclusions from biomedical data. Descriptive statistics summarize the characteristics of a dataset, such as measures of central tendency, variability, and distribution. Inferential statistics enable researchers to make inferences about a population based on sample data, estimating parameters, constructing confidence intervals, and testing hypotheses. Bayesian statistics offers a probabilistic approach to inference, incorporating prior knowledge and uncertainty into the analysis.

Clinical Decision-Making

In clinical settings, statistical methods play a critical role in evidence-based medicine and clinical decision-making. From assessing treatment efficacy and safety to predicting patient outcomes and risk stratification, statistical models inform clinical practice and guide healthcare interventions. For example, risk prediction models based on logistic regression or machine learning algorithms help clinicians estimate the likelihood of disease progression or treatment response for individual patients, facilitating personalized healthcare delivery.

Quality Improvement and Regulatory Compliance

Statistical methods are essential for quality improvement initiatives and regulatory compliance in healthcare and biomedical research. Quality control measures ensure the reliability and reproducibility of data, while statistical process control techniques monitor and maintain the consistency of healthcare processes and laboratory procedures. Regulatory agencies, such as the Food and Drug Administration (FDA) and the European Medicines Agency (EMA), require rigorous statistical analysis and documentation to support drug approval, medical device clearance, and clinical trial validation.

In summary, statistical methods are indispensable tools in biomedical data analysis, providing the foundation for rigorous scientific inquiry, evidence-based decision-making, and quality improvement in healthcare. From experimental design and hypothesis testing to data interpretation and clinical decision support, statistical techniques empower researchers and clinicians to extract actionable insights from biomedical data, advancing scientific knowledge and improving patient outcomes. As biomedical research continues to evolve and generate increasingly complex datasets, the importance of statistical methods in biomedical data analysis will only continue to grow.

Chapter 2: Probability and Statistical Distributions

Probability theory serves as a fundamental tool in bioinformatics for modeling uncertainty, quantifying randomness, and making probabilistic predictions based on biological data. Next, we explore the essential concepts of probability theory and their applications in bioinformatics, covering topics such as probability distributions, conditional probability, Bayes' theorem, and probabilistic modeling.

Probability Distributions

Probability distributions describe the likelihood of different outcomes in a random experiment. In bioinformatics, probability distributions are used to model various biological phenomena, such as the distribution of nucleotide bases in DNA sequences, the expression levels of genes in microarray data, or the frequencies of amino acids in protein sequences. Common probability distributions include the binomial distribution, which describes the number of successes in a fixed number of independent trials with the same probability of success, and the normal distribution, which represents continuous data with a bell-shaped curve.

Conditional Probability

Conditional probability measures the likelihood of an event occurring given that another event has already occurred. In bioinformatics, conditional probability is used to assess the probability of observing certain biological features or patterns given specific conditions.

For example, conditional probability can be applied to predict the probability of finding a specific DNA motif in a genomic sequence given the presence of certain regulatory elements nearby.

Bayes' Theorem

Bayes' theorem is a fundamental concept in probability theory that describes how to update the probability of a hypothesis given new evidence. In bioinformatics, Bayes' theorem is widely used in statistical inference, machine learning, and probabilistic modeling. It provides a framework for combining prior knowledge with observed data to make probabilistic predictions and decisions. Bayes' theorem has applications in sequence alignment, gene expression analysis, protein structure prediction, and many other areas of bioinformatics.

Probabilistic Modeling

Probabilistic modeling involves representing biological phenomena as probabilistic processes and using statistical methods to estimate model parameters and make predictions. In bioinformatics, probabilistic models are used to infer hidden patterns and relationships in biological data, such as hidden Markov models (HMMs) for sequence alignment, Bayesian networks for gene regulatory networks, and probabilistic graphical models for protein-protein interactions. These models capture the inherent uncertainty and variability in biological systems, allowing researchers to account for noise and make more accurate predictions.

Applications in Bioinformatics

Probability theory has numerous applications in bioinformatics, ranging from sequence analysis and phylogenetics to gene expression profiling and functional genomics. For example, in sequence analysis, hidden Markov models are used to align DNA or protein sequences and identify conserved functional regions. In gene expression analysis, probabilistic models are used to identify differentially expressed genes and infer gene regulatory networks. In phylogenetics, probabilistic methods such as maximum likelihood and Bayesian inference are used to reconstruct evolutionary trees and estimate evolutionary distances.

In summary, probability theory provides a powerful framework for modeling uncertainty and making probabilistic predictions in bioinformatics. Understanding basic concepts such as probability distributions, conditional probability, Bayes' theorem, and probabilistic modeling is essential for analyzing biological data, interpreting results, and making informed decisions in bioinformatics research. By leveraging probabilistic methods, researchers can gain insights into the complex and stochastic nature of biological systems, leading to advances in genomics, proteomics, drug discovery, and personalized medicine.

Common Statistical Distributions Used in Biomedical Data Analysis

Statistical distributions play a crucial role in biomedical data analysis by providing a mathematical framework to describe the probability of different outcomes and the distribution of observed data. Next, we explore several common statistical distributions used in biomedical research, including the normal distribution, binomial distribution, Poisson distribution, and exponential distribution. Understanding these distributions is essential for modeling biological phenomena, hypothesis testing, parameter estimation, and making statistical inferences.

Normal Distribution

The normal distribution, also known as the Gaussian distribution, is perhaps the most widely recognized and utilized probability distribution in biomedical data analysis. It is characterized by its bell-shaped curve, with data symmetrically distributed around the mean. Many biological measurements, such as height, weight, blood pressure, and gene expression levels, tend to follow a normal distribution. In hypothesis testing and parameter estimation, researchers often assume normality to make statistical inferences about population parameters.

In R, a popular statistical programming language, the **dnorm, pnorm, qnorm**, and **rnorm** functions are commonly used to work with the normal distribution. For example, to generate random numbers from a normal distribution with a mean of 0 and a standard deviation of 1, the command **rnorm(100, mean = 0, sd = 1)** can be used.

Binomial Distribution

The binomial distribution describes the probability of obtaining a certain number of successes in a fixed number of independent Bernoulli trials. In biomedical research, the binomial distribution is commonly used to model binary outcomes, such as the success or failure of a treatment, the presence or absence of a genetic mutation, or the occurrence of a disease in a population. The binomial distribution is characterized by two parameters: the number of trials (n) and the probability of success (p).

In R, the **dbinom, pbinom, qbinom,** and **rbinom** functions are used to work with the binomial distribution. For example, to calculate the probability of observing exactly 3 successes in 5 trials with a success probability of 0.5, the command **dbinom(3, size = 5, prob = 0.5)** can be used.

Poisson Distribution

The Poisson distribution describes the probability of a given number of events occurring in a fixed interval of time or space, assuming that the events occur independently and at a constant rate. In biomedical research, the Poisson distribution is commonly used to model count data, such as the number of mutations in a gene, the number of cells in a tissue sample, or the number of patients arriving at a hospital emergency room in a given hour.

In R, the **dpois, ppois, qpois,** and **rpois** functions are used to work with the Poisson distribution. For example, to calculate the probability of observing exactly 2 events in a Poisson process with a mean rate of 1 event per unit time, the command **dpois(2, lambda = 1)** can be used.

Exponential Distribution

The exponential distribution describes the probability of the time between consecutive events in a Poisson process, where events occur independently and at a constant rate. In biomedical research, the exponential distribution is commonly used to model waiting times, such as the time until the next mutation occurs, the time until a drug is metabolized by the body, or the time until a patient experiences a relapse after treatment.

In R, the **dexp, pexp, qexp,** and **rexp** functions are used to work with the exponential distribution. For example, to generate random numbers from an exponential distribution with a rate parameter of 0.1, the command **rexp(100, rate = 0.1)** can be used.

In summary, understanding common statistical distributions is essential for biomedical researchers to analyze data, make statistical inferences, and draw meaningful conclusions from their research findings. By leveraging the appropriate statistical distribution for a given dataset or research question, researchers can better model biological phenomena, test hypotheses, and make informed decisions in biomedical research and healthcare. Through the use of statistical distributions, researchers can gain valuable insights into the underlying mechanisms of disease, identify biomarkers, and develop new therapeutic interventions to improve human health and well-being.

Chapter 3: Hypothesis Testing and Statistical Inference

Hypothesis testing is a fundamental concept in bioinformatics, allowing researchers to make statistical inferences about biological phenomena based on observed data. Next, we delve into the principles of hypothesis testing in bioinformatics, exploring the underlying concepts, the steps involved, and the significance of hypothesis testing in scientific research.

Understanding Hypothesis Testing

At its core, hypothesis testing involves comparing observed data to a null hypothesis, which represents the absence of an effect or difference. The null hypothesis typically assumes no relationship, no effect, or no difference between groups or conditions, while the alternative hypothesis proposes a specific relationship, effect, or difference. By collecting data and performing statistical tests, researchers aim to determine whether the observed data provide sufficient evidence to reject the null hypothesis in favor of the alternative hypothesis.

Steps of Hypothesis Testing

The process of hypothesis testing in bioinformatics typically involves several key steps:

Formulating Hypotheses: Researchers begin by defining the null and alternative hypotheses based on the research question or hypothesis being tested. For example, in a gene expression study, the null hypothesis might state that there is no difference in gene expression between two groups, while the alternative hypothesis might propose that there is a significant difference.

Choosing a Statistical Test: The choice of statistical test depends on the nature of the data and the research question. Common statistical tests used in bioinformatics include t-tests, chi-square tests, ANOVA, correlation tests, and non-parametric tests. The appropriate test should be selected based on the assumptions of the test and the characteristics of the data.

Setting Significance Level: Researchers define a significance level, often denoted by α, which represents the probability of rejecting the null hypothesis when it is actually true. The commonly used significance level is $\alpha = 0.05$, indicating a 5% chance of committing a Type I error (false positive).

Collecting Data and Calculating Test Statistic: Researchers collect data and calculate a test statistic based on the chosen statistical test. The test statistic quantifies the difference between the observed data and the null hypothesis, providing a measure of the strength of evidence against the null hypothesis.

Determining Critical Value or P-value: Depending on the chosen statistical test, researchers determine either a critical value or a p-value. The critical value represents the threshold beyond which the null hypothesis is rejected, while the p-value represents the probability of observing the data (or more extreme data) if the null hypothesis is true.

Making a Decision: Researchers compare the test statistic to the critical value or p-value. If the test statistic exceeds the critical value or if the p-value is less than the significance level (α), the null hypothesis is rejected in favor of the alternative hypothesis. Otherwise, the null hypothesis is not rejected.

Deploying Hypothesis Testing Techniques

In bioinformatics, hypothesis testing is commonly deployed using statistical software packages such as R, Python with libraries like SciPy, or specialized bioinformatics software such as Bioconductor. For example, in R, researchers can use functions such as **t.test()** for t-tests, **chisq.test()** for chi-square tests, and **cor.test()** for correlation tests. These functions automatically calculate test statistics, p-values, and other relevant statistics based on the input data.

Importance of Hypothesis Testing in Bioinformatics

Hypothesis testing is essential in bioinformatics for several reasons:

Validation of Research Findings: Hypothesis testing allows researchers to validate research findings by determining whether observed effects or differences are statistically significant.

Identification of Biomarkers: By testing hypotheses about gene expression patterns, genetic variations, or protein interactions, researchers can identify potential biomarkers associated with diseases or biological processes.

Functional Annotation of Genomic Data: Hypothesis testing helps in the functional annotation of genomic data by testing whether genes or gene sets are enriched for specific biological functions or pathways.

Drug Discovery and Target Identification: In drug discovery research, hypothesis testing is used to identify potential drug targets or to assess the efficacy of candidate drugs based on preclinical or clinical data.

Precision Medicine: Hypothesis testing plays a crucial role in precision medicine by identifying molecular signatures associated with individual responses to treatments or disease outcomes.

In summary, hypothesis testing is a fundamental aspect of bioinformatics research, enabling researchers to make evidence-based decisions, validate research findings, and advance our understanding of complex biological systems. By understanding the principles of hypothesis testing and deploying appropriate statistical techniques, bioinformaticians can extract meaningful insights from data, uncover biological mechanisms, and contribute to the development of personalized therapies and precision medicine approaches.

Statistical Inference Methods for Biological Data Analysis

Statistical inference methods play a critical role in analyzing biological data, allowing researchers to draw conclusions and make predictions based on observed data. Next, we explore the principles and applications of statistical inference methods in the context of biological data analysis, covering topics such as hypothesis testing, confidence intervals, regression analysis, and Bayesian inference.

1. Introduction to Statistical Inference

Statistical inference is the process of drawing conclusions about a population based on a sample of data. In biological research, statistical inference is used to make inferences about biological phenomena, such as gene expression patterns, genetic associations, and protein interactions. By applying statistical methods to biological data, researchers can test hypotheses, quantify uncertainty, and make predictions about biological processes.

2. Hypothesis Testing

Hypothesis testing is a fundamental statistical technique used to evaluate hypotheses about population parameters. In biological data analysis, researchers often use hypothesis testing to determine whether observed differences or associations are statistically significant. Common hypothesis tests used in biological research include t-tests, chi-square tests, ANOVA, and non-parametric tests. For example, the t-test can be used to compare the means of two groups of gene expression data, while the chi-square test can assess the association between genotype and phenotype in genetic studies.

To perform hypothesis testing in the command line interface (CLI), researchers can use statistical software packages such as R or Python with libraries like SciPy. For example, in R, the **t.test()** function can be used to perform a t-test:

RCopy code

```
# Load data data <- c(23, 25, 28, 30, 32) # Perform one-sample t-test t.test(data, mu=25)
```

3. Confidence Intervals

Confidence intervals provide a range of plausible values for a population parameter, along with a level of confidence that the true parameter lies within the interval. In biological data analysis, confidence intervals are used to estimate parameters such as means, proportions, and regression coefficients. By calculating confidence intervals, researchers can quantify the precision of their estimates and assess the uncertainty associated with their findings.

To calculate confidence intervals in the CLI, researchers can use statistical functions available in software packages like R or Python. For example, in Python using SciPy:

pythonCopy code

```
import numpy as np from scipy.stats import t # Sample data data = np.array([23, 25, 28, 30, 32]) # Calculate confidence interval for the mean mean = np.mean(data) std_err = np.std(data) / np.sqrt(len(data)) conf_interval = t.interval(0.95, len(data)-1, loc=mean, scale=std_err) print("95% Confidence Interval:", conf_interval)
```

4. Regression Analysis

Regression analysis is a statistical technique used to model the relationship between one or more independent variables and a dependent variable. In biological data analysis, regression analysis is commonly used to assess the association between variables, such as gene expression and clinical outcomes, or to predict biological phenomena based on explanatory variables. Common types of regression analysis include linear regression, logistic regression, and Cox proportional hazards regression.

In the CLI, researchers can perform regression analysis using software packages like R or Python with libraries such as StatsModels or scikit-learn. For example, in R, researchers can use the **lm()** function to fit a linear regression model:

RCopy code

```
# Load data data <- read.csv("data.csv") # Fit linear regression model model <- lm(y ~ x1 + x2, data=data) # Print summary summary(model)
```

5. Bayesian Inference

Bayesian inference is a statistical framework for updating beliefs about parameters or hypotheses based on observed data and prior knowledge. In biological data analysis, Bayesian inference allows researchers to incorporate prior information, such as previous studies or expert knowledge, into their analyses. Bayesian methods can be used for parameter estimation, hypothesis testing, and model comparison, offering a flexible and principled approach to statistical inference.

To perform Bayesian inference in the CLI, researchers can use specialized software packages such as Stan or PyMC3 in Python. For example, in PyMC3, researchers can specify a Bayesian model using a probabilistic programming language and perform inference using Markov chain Monte Carlo (MCMC) sampling:

pythonCopy code

```
import pymc3 as pm # Define Bayesian model with pm.Model() as model: # Define priors mu =
```

```
pm.Normal('mu', mu=0, sigma=10) sigma =
pm.HalfNormal('sigma', sigma=1) # Define likelihood
likelihood = pm.Normal('likelihood', mu=mu,
sigma=sigma, observed=data) # Perform inference
trace = pm.sample(1000, tune=1000) # Summarize
results pm.summary(trace)
```

In summary, statistical inference methods are essential tools for analyzing biological data, allowing researchers to test hypotheses, estimate parameters, and make predictions based on observed data. By understanding and applying statistical techniques such as hypothesis testing, confidence intervals, regression analysis, and Bayesian inference, researchers can gain insights into complex biological processes and contribute to advancements in biomedical research and healthcare.

Chapter 4: Parametric and Nonparametric Tests in Bioinformatics

Parametric statistical tests are a fundamental component of data analysis in various scientific fields, including bioinformatics and biomedical research. These tests are utilized to assess hypotheses and make inferences about population parameters based on sample data. Next, we delve into the principles, applications, and deployment of parametric statistical tests in the context of biological data analysis.

1. Introduction to Parametric Statistical Tests

Parametric statistical tests are based on specific assumptions about the distribution of the data, particularly that it follows a known probability distribution, often the normal distribution. These tests are powerful when their assumptions are met, allowing for precise hypothesis testing and estimation of parameters such as means and variances.

2. t-Tests: Assessing Differences in Means

The t-test is one of the most widely used parametric tests for comparing the means of two groups. In bioinformatics, t-tests are employed to evaluate differences in gene expression levels between experimental conditions or to compare phenotypic traits across populations. The two-sample t-test, for instance, assesses whether the means of two independent groups are significantly different.

To execute a two-sample t-test in the CLI using R, researchers can utilize the **t.test()** function. Here's how it's done:

RCopy code

```
# Load data group1 <- c(10, 12, 15, 14, 11) group2 <- c(13, 16, 18, 17, 19) # Perform two-sample t-test result <- t.test(group1, group2) # View test results print(result)
```

3. Analysis of Variance (ANOVA)

ANOVA is a parametric test used to assess whether there are statistically significant differences in means across three or more groups. In bioinformatics, ANOVA can be applied to analyze gene expression data across multiple experimental conditions or to compare phenotypic traits among several genetic variants.

Executing ANOVA in the CLI with R involves the **aov()** function. Here's a simplified example:

RCopy code

```
# Load data group1 <- c(10, 12, 15, 14, 11) group2 <- c(13, 16, 18, 17, 19) group3 <- c(9, 8, 11, 10, 12) # Perform ANOVA result <- aov(c(group1, group2, group3) ~ c(rep("Group 1", 5), rep("Group 2", 5), rep("Group 3", 5))) # View ANOVA results print(summary(result))
```

4. Analysis of Covariance (ANCOVA)

ANCOVA extends ANOVA by incorporating one or more continuous covariates into the analysis. In bioinformatics, ANCOVA is employed to assess the impact of confounding variables on the relationship between independent and dependent variables. For example, ANCOVA can be used to analyze gene expression data while controlling for factors such as age or gender.

Running ANCOVA in the CLI with R involves the **lm()** function, similar to linear regression. Here's a simplified example:

RCopy code

```
# Load data response <- c(10, 12, 15, 14, 11) covariate <- c(25, 30, 28, 32, 27) group <- factor(c("A", "A", "B", "B", "B")) # Perform ANCOVA result <- lm(response ~ covariate + group) # View ANCOVA results print(summary(result))
```

5. Linear Regression

Linear regression is a parametric statistical method used to model the relationship between one or more independent variables and a continuous dependent variable. In bioinformatics, linear regression is applied to analyze associations between gene expression levels and clinical outcomes or to predict biological traits based on genetic factors.

Conducting linear regression in the CLI with R involves the **lm()** function, as demonstrated in the ANCOVA example above.

Parametric statistical tests are indispensable tools in the analysis of biological data, enabling researchers to make meaningful inferences and discoveries. By understanding the principles and applications of tests such as t-tests, ANOVA, ANCOVA, and linear regression, researchers can effectively analyze complex datasets and uncover valuable insights in bioinformatics and biomedical research.

Nonparametric methods play a crucial role in the analysis of biomedical data, offering robust alternatives to parametric approaches when data distribution assumptions are violated or when dealing with ordinal or nominal data. Next, we explore the principles, applications, and deployment of nonparametric methods in the context of biomedical research.

1. Introduction to Nonparametric Methods

Nonparametric methods are statistical techniques that do not rely on specific assumptions about the underlying distribution of the data. Instead, they make fewer assumptions or use distribution-free approaches, making them versatile and applicable to a wide range of data types and scenarios.

2. Mann-Whitney U Test

The Mann-Whitney U test, also known as the Wilcoxon rank-sum test, is a nonparametric test used to compare the distributions of two independent samples. It assesses whether the medians of the two groups are significantly different, making it suitable for analyzing ordinal or non-normally distributed data.

To perform the Mann-Whitney U test in the CLI using R, the **wilcox.test()** function is utilized. Here's a simplified example:

RCopy code

```
# Load data group1 <- c(10, 12, 15, 14, 11) group2 <- c(13, 16, 18, 17, 19) # Perform Mann-Whitney U test
```

```
result <- wilcox.test(group1, group2) # View test
results print(result)
```

3. Kruskal-Wallis Test

The Kruskal-Wallis test is a nonparametric alternative to one-way ANOVA and is used to compare the distributions of three or more independent groups. It assesses whether there are statistically significant differences in the medians of the groups.

Executing the Kruskal-Wallis test in the CLI with R involves the **kruskal.test()** function. Here's a simplified example:

RCopy code

```
# Load data group1 <- c(10, 12, 15, 14, 11) group2 <-
c(13, 16, 18, 17, 19) group3 <- c(9, 8, 11, 10, 12) #
Perform Kruskal-Wallis test result <-
kruskal.test(list(group1, group2, group3)) # View test
results print(result)
```

4. Spearman's Rank Correlation

Spearman's rank correlation coefficient is a nonparametric measure of the strength and direction of association between two variables. It assesses whether there is a monotonic relationship between the variables, making it suitable for analyzing ordinal or ranked data.

To calculate Spearman's rank correlation coefficient in the CLI using R, the **cor()** function with the method parameter set to "spearman" is used. Here's a simplified example:

RCopy code

```
# Load data x <- c(10, 12, 15, 14, 11) y <- c(13, 16, 18, 17, 19) # Calculate Spearman's rank correlation coefficient result <- cor(x, y, method = "spearman") # View correlation coefficient print(result)
```

5. Wilcoxon Signed-Rank Test

The Wilcoxon signed-rank test is a nonparametric alternative to the paired t-test and is used to compare two related samples. It assesses whether the medians of the differences between paired observations are significantly different from zero.

To perform the Wilcoxon signed-rank test in the CLI using R, the **wilcox.test()** function with the paired parameter set to TRUE is used. Here's a simplified example:

RCopy code

```
# Load data before <- c(10, 12, 15, 14, 11) after <- c(13, 16, 18, 17, 19) # Perform Wilcoxon signed-rank test result <- wilcox.test(before, after, paired = TRUE) # View test results print(result)
```

Nonparametric methods provide valuable tools for analyzing biomedical data, particularly in situations where parametric assumptions are not met or when dealing with ordinal or ranked data. By understanding and applying techniques such as the Mann-Whitney U test, Kruskal-Wallis test, Spearman's rank correlation, and Wilcoxon signed-rank test, researchers can effectively analyze diverse datasets and draw reliable conclusions in biomedical research and bioinformatics.

Chapter 5: Regression Analysis in Bioinformatics

Linear and logistic regression models are fundamental tools in bioinformatics for analyzing relationships between variables and making predictions. Next, we delve into the principles, applications, and deployment of linear and logistic regression models in bioinformatics research.

1. Introduction to Regression Analysis

Regression analysis is a statistical technique used to examine the relationship between a dependent variable and one or more independent variables. Linear regression models are used when the dependent variable is continuous, while logistic regression models are suitable for binary or categorical outcomes.

2. Linear Regression in Bioinformatics

Linear regression is extensively used in bioinformatics to model relationships between variables such as gene expression levels, protein concentrations, and clinical outcomes. In gene expression analysis, linear regression can be employed to identify genes that are differentially expressed across experimental conditions or to predict gene expression based on other variables.

In R, linear regression analysis can be performed using the **lm()** function. Here's an example:

RCopy code

```
# Load data data <- read.csv("gene_expression_data.csv") # Perform linear regression model <- lm(Expression ~ Gene_Length +
```

Treatment, data=data) # Summarize results summary(model)

3. Logistic Regression in Bioinformatics

Logistic regression is widely utilized in bioinformatics for binary classification tasks, such as predicting disease status based on genetic markers or identifying biomarkers associated with drug response. In addition to binary outcomes, logistic regression can also handle ordinal or multinomial response variables.

To perform logistic regression analysis in R, the **glm()** function with the family parameter set to "binomial" is used. Here's an example:

RCopy code

```
# Load data data <- read.csv("clinical_data.csv") # Perform logistic regression model <- glm(Disease_Status ~ Genetic_Marker1 + Genetic_Marker2, family=binomial, data=data) # Summarize results summary(model)
```

4. Regularization Techniques

Regularization techniques such as Ridge regression and Lasso regression are commonly employed to address issues of overfitting and multicollinearity in regression models. These techniques add penalty terms to the regression objective function, encouraging simpler models with fewer parameters.

In R, regularization techniques can be applied using packages such as **glmnet** for Lasso and Ridge regression. Here's an example using Lasso regression:

RCopy code

```
# Load data data <- read.csv("gene_expression_data.csv") # Perform Lasso regression lasso_model <- glmnet(as.matrix(data[, c("Gene1", "Gene2", "Gene3")]), Response, family="binomial", alpha=1) # Plot coefficients plot(lasso_model)
```

5. Model Evaluation and Validation

Once regression models are trained, it is essential to evaluate their performance and validate their predictive accuracy. Techniques such as cross-validation, receiver operating characteristic (ROC) curve analysis, and calculation of performance metrics like accuracy, sensitivity, and specificity are commonly used for model evaluation in bioinformatics.

In R, various packages such as **caret** and **ROCR** can be used for model evaluation and validation. Here's an example of ROC curve analysis:

RCopy code

```
# Load data data <- read.csv("validation_data.csv") # Predict probabilities predicted_probs <- predict(model, newdata=data, type="response") # Calculate ROC curve roc_obj <- roc(data$Actual_Status, predicted_probs) # Plot ROC curve plot(roc_obj)
```

Linear and logistic regression models are powerful tools in bioinformatics for analyzing and predicting biological phenomena. By understanding the principles underlying these models and employing appropriate techniques for model training, evaluation, and validation, researchers can gain valuable insights into complex biological processes and make informed decisions in various biomedical applications.

Regression techniques play a crucial role in predictive modeling within the realm of biomedical data analysis. These techniques are employed to understand and predict relationships between variables, aiding in clinical decision-making, drug discovery, disease prognosis, and various other biomedical applications.

1. Introduction to Regression Techniques

Regression analysis is a statistical method used to model the relationship between one or more independent variables and a dependent variable. In biomedical data analysis, regression techniques are extensively utilized to predict outcomes based on features such as genetic markers, biomarkers, clinical parameters, and environmental factors.

2. Linear Regression

Linear regression is a simple yet powerful technique used to model linear relationships between variables. In biomedical research, linear regression is commonly employed to predict continuous outcomes, such as disease progression, based on independent variables. For example, in pharmacokinetics, linear regression models can predict drug concentrations in the body over time based on dosing regimens and patient characteristics.

In Python, linear regression can be implemented using libraries such as scikit-learn. Here's an example:

```python
pythonCopy code
# Import necessary libraries  from sklearn.linear_model
import LinearRegression import numpy as np # Create
sample data X = np.array([[1, 2], [3, 4], [5, 6]]) y =
np.array([3, 7, 11]) # Initialize and fit the linear
regression model model = LinearRegression().fit(X, y) #
Predict outcomes predictions = model.predict([[7, 8]])
print(predictions)
```

3. Logistic Regression

Logistic regression is a classification technique used to
model the probability of a binary outcome based on one
or more independent variables. In biomedical research,
logistic regression is frequently employed for tasks such
as disease prediction, patient risk stratification, and
identification of biomarkers associated with disease
outcomes.

In R, logistic regression can be performed using the
glm() function. Here's an example:

```r
RCopy code
# Load data data <- read.csv("clinical_data.csv") #
Perform logistic regression model <-
glm(Disease_Status ~ Biomarker1 + Biomarker2,
family=binomial, data=data) # Summarize results
summary(model)
```

4. Ridge and Lasso Regression

Ridge and Lasso regression are regularization
techniques used to address multicollinearity and
overfitting issues in regression models. In biomedical
data analysis, these techniques are valuable for

improving the predictive performance and interpretability of regression models, especially when dealing with high-dimensional data such as gene expression profiles or omics data.

In Python, Ridge and Lasso regression can be implemented using scikit-learn. Here's an example:

pythonCopy code

```
# Import necessary libraries from sklearn.linear_model import Ridge, Lasso import numpy as np # Create sample data X = np.array([[1, 2], [3, 4], [5, 6]]) y = np.array([3, 7, 11]) # Initialize and fit the Ridge regression model ridge_model = Ridge(alpha=0.1).fit(X, y) # Initialize and fit the Lasso regression model lasso_model = Lasso(alpha=0.1).fit(X, y) # Predict outcomes ridge_predictions = ridge_model.predict([[7, 8]]) lasso_predictions = lasso_model.predict([[7, 8]]) print("Ridge predictions:", ridge_predictions) print("Lasso predictions:", lasso_predictions)
```

5. Model Evaluation and Validation

After training regression models, it is essential to evaluate their performance and validate their predictive accuracy. Techniques such as cross-validation, evaluation metrics like mean squared error (MSE) or R-squared (R^2), and visualization methods such as scatter plots or calibration curves are commonly used for model assessment in biomedical data analysis.

In Python, libraries such as scikit-learn provide functions for model evaluation. Here's an example of cross-validation:

```python
# Import necessary libraries
from sklearn.model_selection import cross_val_score
from sklearn.linear_model import LinearRegression
import numpy as np

# Create sample data
X = np.array([[1, 2], [3, 4], [5, 6]])
y = np.array([3, 7, 11])

# Initialize linear regression model
model = LinearRegression()

# Perform cross-validation
scores = cross_val_score(model, X, y, cv=5)
print("Cross-validated scores:", scores)
```

Regression techniques are indispensable tools in predictive modeling for biomedical data analysis. By leveraging linear regression, logistic regression, and regularization techniques such as Ridge and Lasso regression, researchers can build accurate and interpretable models to gain insights into complex biological phenomena and improve clinical decision-making processes. Additionally, proper model evaluation and validation are essential steps to ensure the reliability and robustness of predictive models in biomedical research.

Chapter 6: Survival Analysis and Time-to-Event Data

Survival analysis is a statistical method widely used in bioinformatics to analyze time-to-event data, where the "event" of interest could be anything from the occurrence of a disease, the failure of a biological system, or the death of an organism. This method is particularly valuable in biomedical research for understanding factors influencing survival outcomes, such as disease progression, treatment effectiveness, and patient prognosis. Next, we will explore the fundamentals of survival analysis methods and their applications in bioinformatics.

1. Introduction to Survival Analysis

Survival analysis originated from medical research, particularly in studying patient survival rates. However, its applications have expanded to various fields, including genetics, epidemiology, and clinical trials. At its core, survival analysis focuses on estimating the time until an event of interest occurs and analyzing how various factors influence the timing of these events.

2. Kaplan-Meier Estimator

The Kaplan-Meier estimator is one of the most commonly used techniques in survival analysis for estimating the survival function from censored data. Censoring occurs when the event of interest is not observed for some subjects within the study period, often due to loss to follow-up or the end of the study. The Kaplan-Meier estimator provides a non-parametric estimate of the survival function, allowing researchers to visualize survival curves over time.

In R, the Kaplan-Meier estimator can be implemented using the **survfit()** function from the **survival** package. Here's an example:

RCopy code

```
# Load the survival package library(survival) # Create
survival object with time and event status surv_object
<- Surv(time = time_variable, event = event_status)
# Fit Kaplan-Meier estimator km_fit <-
survfit(surv_object ~ 1) # Plot Kaplan-Meier survival
curve plot(km_fit, xlab = "Time", ylab = "Survival
Probability", main = "Kaplan-Meier Survival Curve")
```

3. Cox Proportional Hazards Model

The Cox proportional hazards model is a widely used regression method in survival analysis for assessing the relationship between covariates and survival outcomes while accounting for censoring. Unlike parametric models, the Cox model does not make assumptions about the underlying distribution of survival times. Instead, it estimates hazard ratios, which represent the relative risk of experiencing the event of interest based on the covariates.

In R, the Cox proportional hazards model can be implemented using the **coxph()** function from the **survival** package. Here's an example:

RCopy code

```
# Fit Cox proportional hazards model cox_model <-
coxph(Surv(time_variable,  event_status)      ~
```

covariate1 + covariate2, data = dataset) # Summarize Cox model results summary(cox_model)

4. Log-Rank Test

The log-rank test is a hypothesis test commonly used in survival analysis to compare the survival curves of two or more groups. It assesses whether there are significant differences in survival times between groups while accounting for censoring. The log-rank test is a non-parametric test that does not assume any specific distribution of survival times.

In R, the log-rank test can be performed using the **survdiff()** function from the **survival** package. Here's an example:

RCopy code

```
# Perform log-rank test logrank_test <- survdiff(Surv(time_variable, event_status) ~ group_variable, data = dataset) # Print log-rank test results print(logrank_test)
```

5. Accelerated Failure Time (AFT) Models

Accelerated failure time (AFT) models are an alternative approach to survival analysis that directly model the relationship between covariates and survival times without assuming proportional hazards. AFT models estimate the effect of covariates on the time scale, providing insights into how these factors accelerate or decelerate the time to the event of interest.

In R, AFT models can be implemented using the **survreg()** function from the **survival** package. Here's an example:

RCopy code

```
# Fit AFT model aft_model <-
survreg(Surv(time_variable, event_status) ~
covariate1 + covariate2, data = dataset, dist =
"weibull") # Summarize AFT model results
summary(aft_model)
```

Survival analysis methods play a crucial role in analyzing time-to-event data in bioinformatics and biomedical research. From estimating survival probabilities with the Kaplan-Meier estimator to assessing the impact of covariates using Cox proportional hazards models, these techniques enable researchers to gain insights into the factors influencing survival outcomes and make informed decisions in clinical and experimental settings. By understanding and applying survival analysis methods, researchers can better understand disease progression, treatment efficacy, and patient prognosis, ultimately advancing our knowledge of complex biological processes.

The Kaplan-Meier estimator and the Cox proportional hazards model are two fundamental techniques in survival analysis, a statistical method extensively used in various fields, including bioinformatics, epidemiology, and clinical research. Both methods play pivotal roles in analyzing time-to-event data, such as patient survival times, disease progression, and treatment outcomes. Next, we will delve into the principles, applications, and implementation of these techniques in bioinformatics.

Kaplan-Meier Estimator

The Kaplan-Meier estimator is a non-parametric method used to estimate the survival function from time-to-event data, where the event of interest could be anything from the occurrence of a disease to the failure of a biological system. This estimator is particularly useful when dealing with censored data, where the event of interest is not observed for all subjects within the study period.

The survival function $S(t)$ estimated by the Kaplan-Meier estimator represents the probability of survival beyond a certain time t. It is calculated as the product of the probabilities of surviving up to each observed event time:

$$S(t) = \prod_{i:\,t_i \leq t} \left(1 - \frac{d_i}{n_i}\right)$$

Where:

t_i is the time of the ith event.

d_i is the number of events (deaths) at time t_i.

n_i is the number of individuals at risk just before time t_i.

The Kaplan-Meier estimator provides a step function that represents the estimated survival probabilities over time. It allows researchers to visualize survival curves and compare survival probabilities between different groups or conditions.

Implementation in R:

In R, the Kaplan-Meier estimator can be implemented using the **survfit()** function from the **survival** package. Here's how you can use it:

RCopy code

```
# Load the survival package library(survival) # Create
survival object with time and event status surv_object
<- Surv(time = time_variable, event = event_status)
# Fit Kaplan-Meier estimator km_fit <-
survfit(surv_object ~ 1) # Plot Kaplan-Meier survival
curve plot(km_fit, xlab = "Time", ylab = "Survival
Probability", main = "Kaplan-Meier Survival Curve")
```

Cox Proportional Hazards Model

The Cox proportional hazards model is a regression
technique used to assess the relationship between
covariates and survival outcomes while accounting for
censoring. Unlike parametric models, the Cox model
does not make assumptions about the underlying
distribution of survival times. Instead, it estimates
hazard ratios, which represent the relative risk of
experiencing the event of interest based on the
covariates.

The Cox model assumes that the hazard function is
proportional across different levels of covariates.
Mathematically, the hazard function $h(\diamond|\diamond)h(t|X)$ for
an individual with covariate values $\diamond X$ at time $\diamond t$ is
given by:

$$h(\diamond|\diamond)=h0(\diamond)\exp(\diamond 1\diamond 1+\diamond 2\diamond 2+...+\diamond\diamond\diamond\diamond)h(t|X)=h0(t)\exp(\beta 1X1+\beta 2X2+...+\beta pXp)$$

Where:

$h0(\diamond)h0(t)$ is the baseline hazard function.

$\diamond 1,\diamond 2,...,\diamond\diamond\beta 1,\beta 2,...,\beta p$ are the coefficients of the
covariates.

$\diamond 1,\diamond 2,...,\diamond\diamond X1,X2,...,Xp$ are the values of the
covariates for the individual.

The Cox model estimates the hazard ratios (exp(**❖❖**)exp(βi)) for each covariate, indicating the multiplicative effect of the covariate on the hazard rate.

Implementation in R:

In R, the Cox proportional hazards model can be implemented using the **coxph()** function from the **survival** package. Here's how you can use it:

RCopy code

```
# Fit Cox proportional hazards model cox_model <- coxph(Surv(time_variable, event_status) ~ covariate1 + covariate2, data = dataset) # Summarize Cox model results summary(cox_model)
```

The Kaplan-Meier estimator and the Cox proportional hazards model are essential tools in survival analysis, allowing researchers to analyze time-to-event data and understand the factors influencing survival outcomes. While the Kaplan-Meier estimator provides non-parametric estimates of survival probabilities and enables visualization of survival curves, the Cox model allows for the assessment of the relationship between covariates and survival outcomes while accounting for censoring. By mastering these techniques, bioinformaticians can gain valuable insights into disease progression, treatment efficacy, and patient prognosis, ultimately advancing our understanding of complex biological processes.

Chapter 7: Experimental Design and Statistical Power

Experimental design is a critical aspect of conducting bioinformatics studies as it directly influences the reliability, validity, and interpretability of the results obtained. It encompasses the planning, execution, and analysis of experiments aimed at answering specific biological questions or testing hypotheses. Next, we will explore the fundamental principles of experimental design in bioinformatics studies, including the key considerations, common designs, and best practices.

Key Considerations in Experimental Design:

Research Question Definition: The first step in experimental design is clearly defining the research question or hypothesis that the study aims to address. This provides a clear focus and guides subsequent decisions regarding experimental setup, data collection, and analysis.

Selection of Experimental Variables: Identifying the variables that will be manipulated or measured in the experiment is essential. These variables may include biological factors (e.g., gene expression levels, protein concentrations) or experimental conditions (e.g., treatment doses, time points).

Control of Confounding Factors: Confounding factors are variables that are correlated with both the independent and dependent variables, potentially leading to spurious associations or biased results. It is crucial to identify and control for these factors during experimental design to ensure the validity of the conclusions drawn.

Sample Size Determination: Adequate sample size is essential for achieving sufficient statistical power to detect meaningful effects or differences. Sample size estimation should consider factors such as the expected effect size, variability, and desired level of statistical significance.

Randomization and Replication: Randomization involves assigning experimental units (e.g., samples, individuals) to treatment groups or conditions randomly to minimize bias and ensure the comparability of groups. Replication involves performing multiple independent observations or measurements for each experimental condition to assess the consistency and reproducibility of the results.

Blocking and Stratification: Blocking involves grouping experimental units based on known sources of variability (e.g., age, gender) to reduce experimental error and increase precision. Stratification involves dividing the sample into homogeneous subgroups based on specific criteria to ensure adequate representation of diverse populations or conditions.

Data Collection and Measurement Techniques: Selecting appropriate data collection and measurement techniques is crucial for obtaining accurate and reliable data. It is essential to use validated methods and standardized protocols to minimize measurement error and bias.

Ethical and Regulatory Considerations: Bioinformatics studies involving human subjects, animal models, or sensitive data must adhere to ethical guidelines and regulatory requirements to ensure the protection of participants' rights and privacy.

Common Experimental Designs in Bioinformatics:

Completely Randomized Design (CRD): In CRD, experimental units are randomly assigned to treatment groups or conditions, with each unit having an equal chance of receiving any treatment. This design is simple but may lack efficiency in certain situations.

Randomized Complete Block Design (RCBD): RCBD involves grouping experimental units into blocks based on a known source of variability (e.g., age, gender) and then randomly assigning treatments within each block. This design allows for more precise estimation of treatment effects by reducing within-block variability.

Factorial Design: Factorial design involves simultaneously studying the effects of multiple factors (e.g., treatments, genotypes) and their interactions on the response variable. This design allows for the investigation of main effects and interaction effects, providing insights into complex biological phenomena.

Time Series Design: Time series design involves collecting repeated measurements of the same variables over time to assess temporal trends, patterns, or responses to interventions. This design is commonly used in longitudinal studies and studies involving dynamic biological processes.

Case-Control Design: Case-control design involves comparing individuals with a specific outcome or condition (cases) to individuals without the outcome or condition (controls) to identify potential risk factors or associations. This design is commonly used in genetic association studies and epidemiological research.

Best Practices in Experimental Design:

Pilot Studies: Conducting pilot studies can help refine experimental protocols, identify potential challenges, and estimate variability, informing sample size calculations and overall study design.

Standardization: Standardizing experimental procedures, protocols, and data collection methods minimizes variability and ensures consistency across experiments and research groups.

Documentation: Maintaining detailed documentation of experimental procedures, protocols, and data is essential for transparency, reproducibility, and sharing findings with the scientific community.

Validation and Quality Control: Regular validation and quality control procedures should be implemented to verify the accuracy, precision, and reliability of experimental assays and measurements.

Statistical Analysis Plan: Developing a detailed statistical analysis plan prior to data collection helps ensure that appropriate statistical methods are applied to address the research question and test hypotheses effectively.

Flexibility: While careful planning is essential, it is also important to remain flexible and adaptable to unexpected challenges or opportunities that may arise during the course of the study.

Statistical power analysis and sample size calculation are essential components of experimental design in bioinformatics and biomedical research. They play a crucial role in ensuring that studies have a high probability of detecting true effects or differences, thus increasing the reliability and validity of the research findings. Next, we will explore the concepts of statistical power, sample size determination, and the importance of conducting power analysis in bioinformatics studies.

Statistical Power:

Statistical power, often denoted as β, represents the probability of correctly rejecting a null hypothesis when it is false. In other words, it measures the ability of a statistical test to detect a true effect or difference, if it exists. A high statistical power indicates a low risk of committing a Type II error (false negative), while a low power increases the likelihood of failing to detect true effects.

Factors Affecting Statistical Power:

Effect Size: The magnitude of the effect or difference being investigated directly influences statistical power. Larger effect sizes are easier to detect and require smaller sample sizes to achieve adequate power.

Significance Level (Type I Error Rate): The significance level, denoted as α, represents the probability of rejecting a null hypothesis when it is true. Commonly used values for α are 0.05 or 0.01, indicating a 5% or 1% chance of committing a Type I error, respectively.

Sample Size: Larger sample sizes generally result in higher statistical power, as they provide more information and reduce sampling variability. However, increasing sample size may not always be feasible due to practical constraints or resource limitations.

Variability: The variability or dispersion of the data also affects statistical power. Higher variability decreases power, as it makes it more challenging to detect true effects against the background of noise.

Sample Size Calculation:

Sample size determination involves estimating the number of participants, samples, or experimental units required to achieve a desired level of statistical power for detecting a specific effect size with a given significance level. Several methods and formulas are available for calculating sample size, depending on the study design, statistical test, and assumptions.

Power Analysis Software: Various software packages, such as G*Power, R packages (e.g., pwr, pROC), and SAS, offer tools for conducting power analysis and sample size calculation. These tools allow researchers to input relevant parameters, such as effect size, significance level, and desired power, and obtain estimates of the required sample size.

bashCopy code

```
# Example R command for sample size calculation using the pwr package install.packages("pwr") library(pwr)
pwr.t.test(d = 0.5, sig.level = 0.05, power = 0.8, type = "two.sample")
```

Formula-Based Approaches: Sample size formulas are available for various study designs and statistical tests, such as t-tests, ANOVA, regression analysis, and chi-square tests. These formulas typically incorporate parameters such as effect size, variability, significance level, and power.

```bash
bashCopy code
# Example formula for sample size calculation in a two-sample t-test n = (2 * (sigma^2) * (Z_alpha/2 + Z_beta)^2) / (delta^2)
```

Simulation Studies: Simulation-based methods involve generating simulated data under different scenarios and assessing the statistical power of various sample sizes. This approach allows researchers to explore the impact of different factors on power and sample size requirements.

Importance of Power Analysis:

Conducting power analysis is crucial for several reasons:

Resource Allocation: Power analysis helps researchers determine the optimal sample size needed to detect meaningful effects, thereby avoiding the collection of excessive or insufficient data.

Publication Bias: Studies with inadequate statistical power are more likely to yield nonsignificant results, which may lead to publication bias, where only significant findings are reported. Adequate power reduces the risk of publication bias by increasing the likelihood of detecting true effects.

Interpretation of Results: Knowledge of statistical power allows researchers to interpret study results more accurately. A nonsignificant result with low power may indicate a lack of evidence against the null hypothesis, rather than evidence of no effect.

Replicability and Generalizability: Studies with adequate power are more likely to produce robust and replicable findings, enhancing the credibility and generalizability of the research findings.

Challenges and Considerations:

Assumptions: Sample size calculations rely on certain assumptions, such as the effect size, variability, and distributional properties of the data. Sensitivity analyses can help assess the robustness of the results to different assumptions.

Complexity: Calculating sample size for complex study designs or analyses may require specialized expertise and statistical software. Collaboration with statisticians or bioinformaticians can facilitate the accurate estimation of sample size.

Dynamic Nature: Sample size calculations may need to be revisited and adjusted as additional information becomes available during the course of the study, such as preliminary results or changes in the research question.

In summary, statistical power analysis and sample size calculation are indispensable tools for designing rigorous and informative bioinformatics studies. By carefully considering factors such as effect size, significance level, and variability, researchers can optimize the design of their experiments and increase the likelihood of obtaining meaningful and interpretable results.

Chapter 8: Bayesian Methods in Bioinformatics

Bayesian statistics is a powerful framework for inference and decision-making that has gained popularity in various fields, including bioinformatics. Unlike classical frequentist statistics, which focuses on estimating parameters and making inferences based on the frequency of observed data, Bayesian statistics approaches uncertainty probabilistically. Next, we will explore the fundamentals of Bayesian statistics and its applications in bioinformatics.

Bayesian Inference:

At the core of Bayesian statistics is Bayes' theorem, which describes how to update our beliefs (or probabilities) about a hypothesis in light of new evidence. The theorem is formulated as follows:

$$P(H|D) = P(D)P(D|H) \times P(H)$$

Where:

$P(H|D)$ is the posterior probability of hypothesis H given the data D.

$P(D|H)$ is the likelihood of observing data D given hypothesis H.

$P(H)$ is the prior probability of hypothesis H.

$P(D)$ is the probability of observing data D (also known as the marginal likelihood or evidence).

Bayesian inference involves updating our prior beliefs about a hypothesis using observed data to obtain the posterior distribution of the hypothesis. This posterior distribution encapsulates our updated uncertainty about the hypothesis given the data.

Bayesian Modeling:

In Bayesian statistics, parameters of interest are treated as random variables with associated prior distributions. These prior distributions reflect our beliefs about the parameters before observing any data. By combining the prior distributions with the likelihood function, which quantifies how likely the observed data are under different parameter values, we obtain the posterior distribution of the parameters.

Bayesian modeling allows for the incorporation of prior knowledge and uncertainty into the analysis, making it particularly useful in situations where data are limited or noisy. Additionally, Bayesian models provide a natural framework for hierarchical modeling, where parameters at different levels of a system are modeled jointly.

Markov Chain Monte Carlo (MCMC):

One of the key challenges in Bayesian statistics is computing the posterior distribution, especially in complex models where analytical solutions are intractable. Markov Chain Monte Carlo (MCMC) methods provide a powerful class of algorithms for sampling from the posterior distribution.

MCMC algorithms, such as the Metropolis-Hastings algorithm and Gibbs sampling, generate a sequence of samples from the posterior distribution by constructing a Markov chain that converges to the desired distribution. These samples can then be used to estimate summary statistics, construct credible intervals, and make predictions.

Applications in Bioinformatics:

Bayesian statistics has found numerous applications in bioinformatics, ranging from sequence analysis to network modeling and systems biology. Some common applications include:

Gene Expression Analysis: Bayesian methods are used to model gene expression data, identify differentially expressed genes, and infer gene regulatory networks.

Variant Calling: Bayesian models are employed in variant calling from next-generation sequencing data, where they integrate information from multiple sources to accurately identify genetic variants.

Phylogenetics: Bayesian phylogenetic methods are widely used to reconstruct evolutionary trees and estimate divergence times, mutation rates, and ancestral states.

Protein Structure Prediction: Bayesian statistical methods are applied to predict protein structures from amino acid sequences, incorporating information from known protein structures and physical principles.

Drug Discovery: Bayesian modeling is used in drug discovery and pharmacogenomics to predict drug-target interactions, identify potential drug candidates, and optimize drug dosage regimens.

:

In summary, Bayesian statistics provides a flexible and principled framework for inference and decision-making in bioinformatics. By explicitly modeling uncertainty and incorporating prior knowledge, Bayesian methods offer a powerful approach to analyzing complex biological data and making reliable predictions. As bioinformatics continues to advance, Bayesian statistics is expected to play an increasingly important role in tackling challenging problems and extracting meaningful insights from biological data.

Bayesian inference is a powerful statistical framework that provides a systematic approach for updating beliefs and making predictions in the presence of uncertainty. In the context of biomedical data analysis, Bayesian methods offer several advantages, including the ability to incorporate prior knowledge, handle complex models, and quantify uncertainty. This chapter explores the fundamentals of Bayesian inference and its applications in various areas of biomedical data analysis.

Fundamentals of Bayesian Inference:

At the core of Bayesian inference is Bayes' theorem, which provides a principled way to update our beliefs about a hypothesis or parameter given observed data. The theorem is expressed as:

$$P(\vartheta|D) = \frac{P(D|\vartheta) \times P(\vartheta)}{P(D)}$$

Where:

$P(\vartheta|D)$ is the posterior probability of the parameter ϑ given the data D.

$P(D|\vartheta)$ is the likelihood of observing the data D given the parameter ϑ.

$P(\vartheta)$ is the prior probability distribution of the parameter ϑ.

$P(D)$ is the marginal likelihood, also known as the evidence or the probability of observing the data.

Bayesian inference involves updating the prior distribution of a parameter based on observed data to obtain the posterior distribution. This posterior distribution encapsulates our updated beliefs about the parameter after considering the data.

Bayesian Modeling:

In Bayesian modeling, parameters are treated as random variables with associated prior distributions. These prior distributions represent our beliefs about the parameters before observing any data. By combining the prior distributions with the likelihood function, which quantifies the probability of observing the data given different parameter values, we obtain the posterior distribution of the parameters.

Bayesian modeling allows for the incorporation of prior knowledge and uncertainty into the analysis. It also provides a framework for hierarchical modeling, where parameters at different levels of a system are modeled jointly. This flexibility makes Bayesian modeling particularly well-suited for complex biomedical data analysis tasks.

Markov Chain Monte Carlo (MCMC) Methods:

In many cases, computing the posterior distribution analytically is intractable, especially for complex models. Markov Chain Monte Carlo (MCMC) methods offer a practical approach for sampling from the posterior distribution.

MCMC algorithms, such as the Metropolis-Hastings algorithm and Gibbs sampling, generate samples from the posterior distribution by constructing a Markov chain that converges to the desired distribution. These samples can then be used to estimate summary statistics, construct credible intervals, and make predictions.

Applications in Biomedical Data Analysis:

Bayesian inference has numerous applications in biomedical data analysis, spanning various domains, including genomics, proteomics, medical imaging, and clinical trials. Some common applications include:

Genomic Data Analysis: Bayesian methods are used for gene expression analysis, variant calling, genome-wide association studies (GWAS), and gene regulatory network inference.

Medical Imaging: Bayesian techniques are employed for image reconstruction, segmentation, registration, and classification in medical imaging applications such as MRI, CT, and PET.

Clinical Trials: Bayesian methods are used for trial design, patient recruitment, treatment efficacy evaluation, and personalized medicine in clinical trials.

Disease Modeling: Bayesian modeling is applied to epidemiological studies, disease progression modeling, outbreak prediction, and risk assessment.

Drug Discovery: Bayesian approaches are used for drug target identification, pharmacokinetic modeling, dose-finding studies, and drug response prediction.

In summary, Bayesian inference provides a flexible and principled framework for analyzing biomedical data and making informed decisions under uncertainty. By integrating prior knowledge, handling complex models, and quantifying uncertainty, Bayesian methods offer valuable insights into biological processes and enable more accurate predictions and interpretations of biomedical data. As biomedical research continues to advance, Bayesian inference is expected to play an increasingly important role in addressing complex challenges and driving innovation in healthcare and life sciences.

Chapter 9: Advanced Topics in Biostatistics: Meta-analysis and Multilevel Modeling

Meta-analysis is a powerful statistical technique used to synthesize and integrate findings from multiple independent studies on a particular research question. It provides a systematic approach to combine data from different sources, increasing the statistical power and generalizability of results. Next, we will explore the fundamentals of meta-analysis, various methods for conducting meta-analyses, and their applications in biomedical research.

Fundamentals of Meta-analysis:

Meta-analysis involves the systematic review, quantitative synthesis, and analysis of data from multiple studies to derive overall estimates of treatment effects, associations, or outcomes. It aims to provide a more comprehensive understanding of the research question by pooling information across studies and assessing the consistency and robustness of findings.

Types of Meta-analysis:

There are several types of meta-analysis, depending on the nature of the data and the research question:

Fixed-effects Meta-analysis: Assumes that all studies share a common effect size and estimates a single overall effect size across studies.

Random-effects Meta-analysis: Allows for heterogeneity among studies by assuming that each study has its own true effect size, drawn from a distribution of effects.

Individual Participant Data (IPD) Meta-analysis: Combines raw data from individual participants across studies, allowing for more detailed analysis and exploration of heterogeneity.

Steps in Conducting Meta-analysis:

Formulating Research Question: Clearly define the research question and specify the inclusion criteria for selecting studies.

Literature Search: Conduct a comprehensive literature search to identify relevant studies using databases such as PubMed, Web of Science, and Embase.

Study Selection: Screen and select studies based on predefined criteria, such as study design, population characteristics, and outcome measures.

Data Extraction: Extract relevant data from selected studies, including study characteristics, sample sizes, effect estimates, and measures of variability.

Statistical Analysis: Perform statistical analysis to estimate overall effect sizes, assess heterogeneity among studies, and conduct sensitivity analyses.

Interpretation and Reporting: Interpret the results of meta-analysis, discuss the implications for practice and policy, and report findings following guidelines such as PRISMA (Preferred Reporting Items for Systematic Reviews and Meta-Analyses).

Meta-analysis Techniques:

Effect Size Estimation: Common effect size measures include odds ratios, risk ratios, hazard ratios, mean differences, and standardized mean differences.

Heterogeneity Assessment: Quantify and assess heterogeneity among studies using measures such as Cochran's Q statistic and the I^2 statistic.

Publication Bias Detection: Detect and adjust for publication bias using funnel plots, Egger's regression test, and trim-and-fill methods.

Subgroup Analysis: Explore sources of heterogeneity by conducting subgroup analyses based on study characteristics, participant characteristics, or methodological factors.

Sensitivity Analysis: Evaluate the robustness of meta-analysis results by conducting sensitivity analyses, such as excluding low-quality studies or studies with outlier effect sizes.

Applications of Meta-analysis in Biomedical Research:

Meta-analysis has diverse applications across various domains of biomedical research, including:

Clinical Trials: Synthesizing evidence from multiple clinical trials to evaluate the efficacy and safety of interventions and inform clinical practice guidelines.

Epidemiological Studies: Combining data from epidemiological studies to estimate the association between risk factors and disease outcomes.

Genetic Studies: Integrating data from genome-wide association studies (GWAS) to identify genetic variants associated with complex diseases.

Diagnostic Accuracy Studies: Pooling data from diagnostic accuracy studies to assess the performance of diagnostic tests and biomarkers.

Health Services Research: Synthesizing data from health services research studies to evaluate the effectiveness and cost-effectiveness of healthcare interventions and policies.

Challenges and Considerations:

Despite its benefits, meta-analysis also presents several challenges and considerations, including:

Publication Bias: The tendency for studies with positive or significant results to be published, leading to overestimation of treatment effects.

Heterogeneity: Variability in study populations, methodologies, and outcome measures across studies, which can affect the validity and interpretation of meta-analysis results.

Data Quality: Reliance on published data, which may be subject to reporting bias, incomplete reporting, or methodological limitations.

Study Validity: Assessing the quality and validity of included studies to ensure that meta-analysis results are based on reliable evidence.

Meta-analysis is a valuable tool for synthesizing and integrating data from multiple studies in biomedical research. By combining evidence from diverse sources, meta-analysis provides more robust estimates of treatment effects, associations, and outcomes, facilitating evidence-based decision-making and advancing knowledge in the field. However, careful planning, rigorous methodology, and critical appraisal of evidence are essential to ensure the validity and reliability of meta-analysis findings.

Multilevel modeling, also known as hierarchical linear

modeling or mixed-effects modeling, is a powerful statistical technique used to analyze complex data structures characterized by nested or hierarchical levels. In the context of bioinformatics and biomedical research, multilevel modeling offers a flexible framework for analyzing data with multiple levels of variation, such as longitudinal studies, clustered or nested data, and hierarchical data structures. Next, we will explore the principles of multilevel modeling, its applications in analyzing biological data, and practical considerations for implementing multilevel models using statistical software.

Principles of Multilevel Modeling:

At its core, multilevel modeling recognizes that data points within the same group or cluster (e.g., patients within hospitals, gene expression measurements within individuals) are likely to be more similar to each other than to data points from different groups. Therefore, traditional linear models that assume independence of observations may not adequately account for the hierarchical structure of the data. Multilevel modeling addresses this issue by explicitly modeling the nested nature of the data, allowing for the estimation of both within-group and between-group variation.

Key Concepts in Multilevel Modeling:

Level Structure: Multilevel models consist of multiple levels of analysis, with individual data points nested within higher-level units (e.g., patients within hospitals, genes within individuals). These hierarchical structures can be represented mathematically using nested random effects.

Random Effects: Random effects represent the variation between higher-level units (e.g., hospitals, individuals) and are typically assumed to be normally distributed with a mean of zero. They capture unobserved heterogeneity between groups and allow for the estimation of group-level effects.

Fixed Effects: Fixed effects represent the average effect across all levels of the hierarchy and are typically used to estimate the effects of predictors or covariates on the outcome variable.

Variance Components: Multilevel models estimate variance components at each level of the hierarchy, quantifying the proportion of total variance attributable to within-group variation and between-group variation.

Applications of Multilevel Modeling in Bioinformatics:

Longitudinal Data Analysis: Multilevel models are well-suited for analyzing longitudinal data with repeated measurements nested within individuals over time. They allow for the estimation of individual trajectories and the identification of predictors of change over time.

Gene Expression Studies: Multilevel models can be used to analyze gene expression data with multiple levels of variation, such as gene expression measurements nested within individuals or tissues. They enable the identification of differentially expressed genes while accounting for the correlation structure of the data.

Pharmacokinetic Modeling: Multilevel models are commonly used in pharmacokinetic studies to analyze drug concentration data collected from multiple individuals over time. They allow for the estimation of individual-specific parameters while accounting for between-subject variability.

Ecological Studies: Multilevel models are valuable for analyzing ecological data with nested or hierarchical structures, such as species nested within habitats or sampling sites. They facilitate the estimation of species-level effects while accounting for the correlation structure of the data.

Practical Considerations for Implementing Multilevel Models:

Model Specification: Careful consideration should be given to the specification of the multilevel model, including the selection of appropriate random effects, fixed effects, and covariates. Model building techniques such as stepwise selection or information criteria can help identify the most parsimonious model.

Software Implementation: Multilevel models can be implemented using a variety of statistical software packages, including R, SAS, and Stata. In R, the **lme4** package provides a flexible framework for fitting linear mixed-effects models, while the **nlme** package offers functions for fitting nonlinear mixed-effects models.

Model Assessment: Diagnostic checks, such as residual plots, likelihood ratio tests, and goodness-of-fit statistics, should be used to assess the adequacy of the multilevel model and identify potential violations of model assumptions.

Interpretation of Results: Interpretation of multilevel model results should take into account the estimated fixed effects, random effects, and variance components at each level of the hierarchy. Post hoc analyses, such as pairwise comparisons or interaction tests, can further elucidate the relationships between variables.

Multilevel modeling is a versatile statistical technique for analyzing complex biological data with nested or hierarchical structures. By explicitly modeling the nested nature of the data, multilevel models provide valuable insights into within-group and between-group variation, allowing researchers to better understand the underlying processes driving biological phenomena. When applied appropriately and interpreted judiciously, multilevel modeling can enhance the validity and robustness of statistical analyses in bioinformatics and biomedical research.

Chapter 10: Practical Applications and Case Studies in Bioinformatics Biostatistics

Biostatistical methods play a crucial role in bioinformatics, providing the analytical framework necessary for extracting meaningful insights from complex biological data. Next, we will explore the real-world applications of biostatistical methods in bioinformatics across various domains, including genomics, proteomics, transcriptomics, and metabolomics. Through a combination of theoretical principles and practical examples, we will demonstrate how biostatistical methods are deployed to address key research questions and drive discoveries in the field of bioinformatics.

Genomics:

In genomics, biostatistical methods are used to analyze large-scale genomic data, such as DNA sequencing data, to elucidate the genetic basis of complex traits and diseases. One common application is genome-wide association studies (GWAS), where biostatistical techniques are employed to identify genetic variants associated with diseases or phenotypes. The **plink** command-line tool is frequently used to perform association analyses by computing statistical tests, such as chi-square tests or logistic regression, to assess the association between genetic variants and phenotypes.

Another important application of biostatistics in genomics is differential gene expression analysis. RNA-sequencing (RNA-seq) data is analyzed using methods such as edgeR or DESeq2, which utilize negative binomial models to identify genes that are differentially expressed between conditions. The **edgeR** and **DESeq2** R packages provide functions for fitting these models and estimating fold changes and statistical significance.

Proteomics:

In proteomics, biostatistical methods are employed to analyze mass spectrometry data and identify differentially expressed proteins between experimental conditions. Quantitative proteomics experiments generate large datasets consisting of protein abundance measurements across samples, which require sophisticated statistical techniques for analysis. Tools such as MaxQuant and Proteome Discoverer utilize algorithms based on statistical modeling to infer protein abundance and detect differential expression.

Transcriptomics:

Transcriptomics studies aim to characterize gene expression patterns across different biological conditions, tissues, or developmental stages. Biostatistical methods are used to analyze transcriptome data generated from microarray or RNA-seq experiments. Differential gene expression analysis, pathway enrichment analysis, and co-expression network analysis are common applications of biostatistics in transcriptomics. Tools like limma and WGCNA provide functions for performing these analyses and visualizing the results.

Metabolomics:

Metabolomics involves the comprehensive analysis of small molecules (metabolites) present in biological samples. Biostatistical methods are utilized to identify metabolites associated with physiological processes, diseases, or environmental exposures. Multivariate statistical techniques such as principal component analysis (PCA) and partial least squares-discriminant analysis (PLS-DA) are commonly used to analyze metabolomics data and identify patterns or clusters of metabolites associated with different experimental conditions.

Clinical Trials:

Biostatistical methods play a critical role in the design, analysis, and interpretation of clinical trials in bioinformatics and biomedical research. Randomized controlled trials (RCTs) are used to evaluate the efficacy and safety of interventions, such as drugs or medical devices. Biostatistical techniques such as randomization, sample size calculation, and statistical hypothesis testing are employed to design RCTs and analyze the resulting data. The **randomize** command-line tool is often used to generate random allocation sequences for assigning participants to treatment groups in clinical trials.

Machine Learning:

In addition to traditional statistical methods, machine learning techniques are increasingly being applied to analyze biological data in bioinformatics. Supervised learning algorithms such as support vector machines (SVM), random forests, and neural networks are used for classification and prediction tasks, such as disease diagnosis and outcome prediction. Unsupervised learning algorithms like hierarchical clustering and k-means clustering are employed for data exploration and clustering analysis.

Biostatistical methods play a central role in bioinformatics, enabling researchers to extract meaningful insights from complex biological data. From genomics and proteomics to transcriptomics and metabolomics, biostatistics provides the analytical framework necessary for analyzing high-dimensional datasets and identifying patterns, associations, and relationships. By leveraging biostatistical methods, researchers can gain a deeper understanding of biological systems and drive discoveries that have implications for human health and disease.

Biostatistics plays a crucial role in biomedical research, providing the tools and techniques necessary for analyzing complex data and drawing meaningful conclusions. Next, we will explore several case studies that illustrate the diverse applications of biostatistics in biomedical research. Through these examples, we will demonstrate how biostatistical methods are deployed to address various research questions, from investigating the genetic basis of diseases to evaluating the efficacy of medical interventions.

Case Study 1: Genome-Wide Association Study (GWAS) for Complex Diseases

One of the most common applications of biostatistics in biomedical research is genome-wide association studies

(GWAS), which aim to identify genetic variants associated with complex diseases. In this case study, researchers are investigating the genetic basis of type 2 diabetes mellitus (T2DM) using data from a large-scale GWAS.

The researchers first perform quality control (QC) on the genotyping data to remove low-quality samples and markers using tools such as **plink**. Next, they conduct association analysis using logistic regression to test for association between genetic variants and T2DM status. The **plink --assoc** command is used to perform association analysis, and the results are corrected for multiple testing using methods such as Bonferroni correction or false discovery rate (FDR) correction.

The analysis identifies several genetic loci significantly associated with T2DM, providing insights into the genetic architecture of the disease and potential targets for further investigation.

Case Study 2: Clinical Trial for a Novel Drug

In this case study, researchers are conducting a randomized controlled trial (RCT) to evaluate the efficacy of a novel drug for the treatment of hypertension. The RCT involves recruiting participants with hypertension and randomly assigning them to receive either the experimental drug or a placebo.

Biostatistical methods are used to design the trial, including sample size calculation to ensure adequate statistical power to detect a clinically meaningful difference between treatment groups. The **sampleSize** R package is often used to perform sample size calculation based on parameters such as expected effect size, significance level, and power.

During the trial, biostatistical techniques are employed to analyze the data, including descriptive statistics to summarize baseline characteristics of participants and inferential statistics to compare outcomes between treatment groups. For continuous outcomes such as blood pressure, analysis of covariance (ANCOVA) is commonly used to adjust for baseline covariates and assess treatment effects.

The results of the trial are analyzed using appropriate statistical tests, such as t-tests or chi-square tests for continuous and categorical outcomes, respectively. The findings of the trial provide evidence regarding the efficacy and safety of the novel drug, informing clinical practice and future research directions.

Case Study 3: Meta-Analysis of Clinical Trials

Meta-analysis is a powerful technique used to synthesize evidence from multiple studies to provide more robust estimates of treatment effects. In this case study, researchers are conducting a meta-analysis to evaluate the effectiveness of a particular intervention for smoking cessation across multiple randomized controlled trials.

The researchers first identify relevant studies through systematic literature review and assess their quality using criteria such as study design, sample size, and risk of bias. Data from individual studies are extracted and combined using statistical techniques such as fixed-effects or random-effects meta-analysis.

The **meta** package in R is commonly used to perform meta-analysis, providing functions for data synthesis and estimation of overall treatment effects. Heterogeneity between studies is assessed using statistical tests such as Cochran's Q test and I^2 statistic.

The results of the meta-analysis provide a comprehensive summary of the effectiveness of the intervention across multiple studies, with implications for clinical practice and public health policy.

These case studies illustrate the diverse applications of biostatistics in biomedical research, from identifying genetic variants associated with diseases to evaluating the efficacy of medical interventions. By leveraging biostatistical methods, researchers can extract meaningful insights from complex data and make informed decisions that advance our understanding of human health and disease.

Conclusion

In summary, the "Bioinformatics Algorithms, Coding, Data Science, and Biostatistics" book bundle offers a comprehensive exploration of key topics at the intersection of biology, computer science, statistics, and data science. Across four volumes, readers are introduced to fundamental concepts, techniques, and methodologies that are essential for understanding and advancing research in bioinformatics.

In Book 1, "Bioinformatics Basics: An Introduction to Algorithms and Concepts," readers are provided with a solid foundation in bioinformatics, learning about the principles of sequence analysis, alignment algorithms, genetic variation, and more. This book serves as a gateway to the field, offering insights into the computational methods that underpin much of modern biological research.

Book 2, "Coding in Bioinformatics: From Scripting to Advanced Applications," takes readers on a journey through the practical implementation of bioinformatics algorithms and techniques. From scripting languages like Python and R to advanced applications in data manipulation, visualization, and machine learning, readers gain hands-on experience in coding for bioinformatics.

In Book 3, "Exploring Data Science in Bioinformatics: Techniques and Tools for Analysis," the focus shifts to the burgeoning field of data science and its applications in bioinformatics. Readers learn about exploratory data analysis, statistical inference, machine learning, and data visualization, equipping them with the tools needed to extract meaningful insights from biological data.

Finally, Book 4, "Mastering Biostatistics in Bioinformatics: Advanced Methods and Applications," delves into the intricacies of biostatistics and its role in bioinformatics research. Readers explore advanced statistical methods, survival analysis, meta-analysis, and more, gaining a deeper understanding of how statistical techniques can be applied to address complex biological questions.

Collectively, this book bundle provides a comprehensive and interdisciplinary approach to bioinformatics, equipping readers with the knowledge and skills needed to tackle real-world challenges in biological research. Whether you are a novice seeking an introduction to the field or an experienced researcher looking to expand your skill set, the "Bioinformatics Algorithms, Coding, Data Science, and Biostatistics" bundle offers something for everyone, fostering a deeper understanding of the intersection between biology and computational science.